ENCYCLOPEDIA OF
REPTILES &
AMPHIBIANS

ENCYCLOPEDIA OF
REPTILES &
AMPHIBIANS

CONSULTANT EDITORS
Dr Harold G. Cogger
Australian Museum, Sydney, Australia

&

Dr Richard G. Zweifel
American Museum of Natural History, New York, USA

ILLUSTRATIONS BY
Dr David Kirshner

NATURAL WORLD

Published in the United States by
Academic Press, A Division of Harcourt Brace & Company
525 B Street, Suite 1900, San Diego, CA 92101-4495, USA
Distributed worldwide exclusively by Academic Press
(except in Australia and New Zealand)

Senior Editor, Life Sciences: Charles R. Crumly Ph.D.

Conceived and produced by Weldon Owen Pty Limited
59 Victoria Street, McMahons Point, NSW 2060, Australia
A member of the Weldon Owen Group of Companies
Sydney • San Francisco

First published in 1992
Second edition 1998
Copyright © Weldon Owen Pty Limited 1998

Publisher: Sheena Coupe
Associate Publisher: Lynn Humphries
Project Editors: Jenni Bruce; Helen Cooney
Editorial Assistant: Veronica Hilton
Captions: David Kirshner; Terence Lindsey
Index: Garry Cousins
Designers: Denese Cunningham; Sue Rawkins
Picture Research: Esther Beaton; Annette Crueger
Production Manager: Caroline Webber
Production Assistant: Kylie Lawson
Vice President International Sales: Stuart Laurence

Co-ordination of scientific and editorial contributors by
Linda Gibson, Project Manager, Australian Museum Business Services

ISBN 0-12-178560-2

A catalog record for this book is available from
the Library of Congress, Washington, DC.

Printed by Kyodo Printing Co. (Singapore) Pte Ltd
Printed in Singapore

A WELDON OWEN PRODUCTION

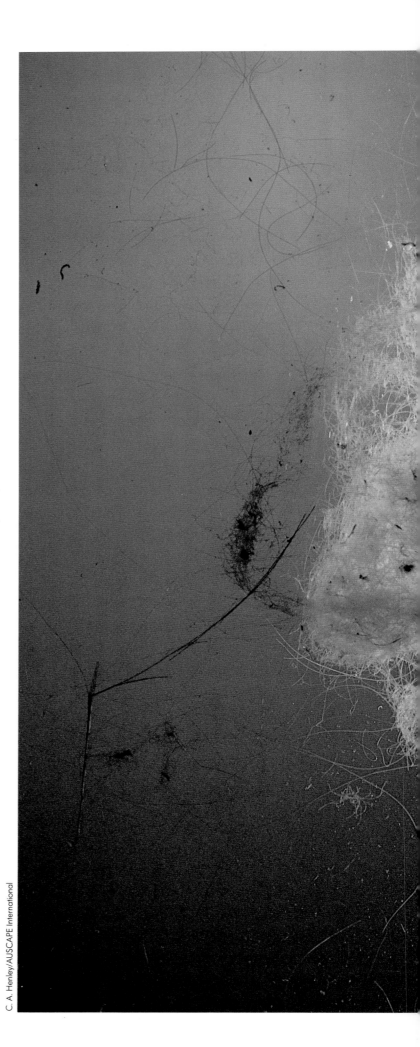

Endpapers: Bathed in golden light, green turtles *Chelonia mydas* come ashore at dusk to lay their eggs. Photo by D. Parer and E. Parer-Cook/AUSCAPE International
Page 1: The eyes of the Mediterranean chameleon *Chamaeleo chameleon* rotate independently, allowing it to scan in all directions for prey and possible danger. Photo by Stephen Dalton/NHPA
Pages 2–3: Turtles sunning themselves on a log. Photo by Kim Heacox/The Photo Library, Sydney
Pages 4–5: A female spotted grass frog *Limnodynastes tasmaniensis* with her egg mass.
Page 7: The flying or parachute gecko *Ptychozoon homocephalum* takes the easy way down from the treetops using folds of skin on the side of its body and between its toes. Photo by Stephen Dalton/NHPA
Pages 10–11: A marine turtle uses its huge, wing-like flippers to propel itself through the water.
Pages 12–13: Galapagos giant tortoises *Chelonoidis elephantopus* grazing, Isabela Island. Photo by Tui De Roy/AUSCAPE International
Pages 50–51: Perhaps the best known of all frogs, the common frog *Rana temporaria* of Europe leaps into a pond. Photo by Stephen Dalton
Pages 106–107: A marine iguana *Amblyrhynchus cristatus* returns to the shore after grazing on algae. Photo by Mark Jones/AUSCAPE International

C. A. Henley/AUSCAPE International

CONSULTANT EDITORS

DR HAROLD G. COGGER
John Evans Memorial Fellow,
Australian Museum, Sydney, and
Conjoint Professor,
Faculty of Science and Mathematics,
University of Newcastle, Australia

DR RICHARD G. ZWEIFEL
Curator Emeritus,
Department of Herpetology and Ichthyology,
American Museum of Natural History,
New York, USA

CONTRIBUTORS

DR AARON M. BAUER
Professor of Biology,
Villanova University,
Pennsylvania, USA

DR CHARLES C. CARPENTER
Professor Emeritus of Zoology,
University of Oklahoma, and
Curator Emeritus of Reptiles and Amphibians,
Oklahoma Museum of Natural History, USA

DR WILLIAM E. DUELLMAN
Curator Emeritus, Division of Herpetology,
Museum of Natural History, and
Professor Emeritus, Department of Systematics and Ecology,
University of Kansas, USA

DR CARL GANS
Adjunct Professor of Zoology,
Department of Zoology,
University of Texas at Austin, USA

DR BRIAN GROOMBRIDGE
Biodiversity Assessment Programme,
World Conservation Monitoring Centre,
Cambridge, England

DR HAROLD HEATWOLE
Professor, Department of Zoology,
North Carolina State University,
Raleigh, USA

DR BENEDETTO LANZA
formerly Director of Zoological Museum "La Specola", and
Professor of General Biology,
University of Florence, Italy

DR WILLIAM E. MAGNUSSON
Research Scientist,
Instituto Nacional de Pesquisas da Amazonia, Brazil

DR DONALD G. NEWMAN
Science and Research,
Science, Technology and Information Services,
Department of Conservation, Wellington,
New Zealand

DR ANNAMARIA NISTRI
Graduate Technician,
Zoological Museum "La Specola",
University of Florence, Italy

DR RONALD A. NUSSBAUM
Professor of Biology, and
Curator of Amphibians and Reptiles,
Museum of Zoology,
University of Michigan, USA

DR FRITZ JURGEN OBST
Curator of Herpetology and Vice-Director,
State Museum of Zoology,
Dresden, Germany

DR OLIVIER C. RIEPPEL
Curator of Fossil Amphibians and Reptiles,
Department of Geology,
Field Museum of Natural History,
Chicago, USA

DR JAY M. SAVAGE
Professor of Biology, University of Miami,
Coral Gables, Florida, USA

DR RICHARD SHINE
Professor in Evolutionary Biology,
University of Sydney, Australia

DR STEFANO VANNI
Graduate Technician,
Zoological Museum "La Specola",
University of Florence, Italy

CONTENTS

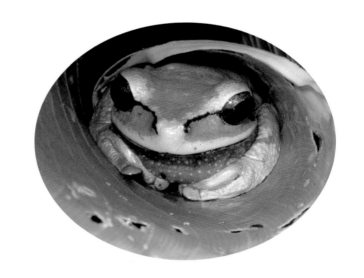

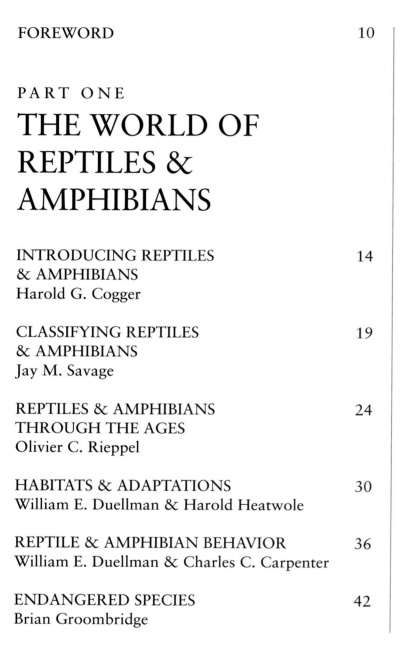

PART THREE
KINDS OF REPTILES

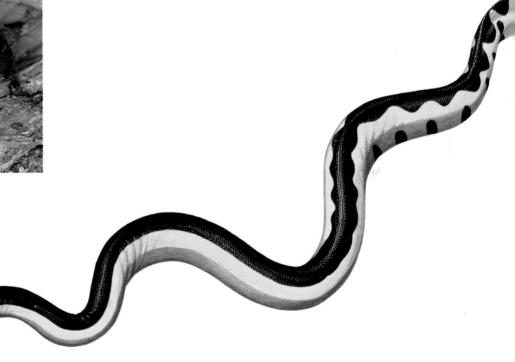

FOREWORD

Dinosaurs are unquestionably the reptiles that most strongly capture our attention and imagination. And little wonder, because these giants dominated the land for more than 100 million years, until their rather sudden extinction about 65 million years ago. Fortunately, although not as conspicuous, other major groups of reptiles have survived until today—as have many amphibians, whose ancestors were the first backboned creatures to colonize the land.

The sheer variety of living reptiles and amphibians is amazing. Considering the constraints of their limited body forms and cold-bloodedness, they have an astonishing array of lifestyles and survival strategies. Frogs, for example, exploit two very different domains—that of predators as adults and herbivores as tadpoles. Snakes dwell in a range of environments, from subterranean to arboreal to aquatic, including the seas. And the ways in which reptiles and amphibians mate and move are equally broad, with parental care, courtship, territoriality, and migration rivaling the birds and mammals. Clearly, though, we have much more to learn about these wonderful animals, including their ecological importance.

It is also clear that we must manage the effects of our own human species on the environment if we are to help these creatures and benefit from them. Destruction of habitats, degradation of the environment by pollution, introduction of alien species, and direct taking continue to threaten the existence of many species. The imminent extinction of some species has, however, been halted by recent conservation initiatives. Several species of crocodiles, and the people who harvest them, are benefiting from the tactics of sustainable utilization, and new international accords are giving greater protection to the marine turtles. Yet we continue to receive disturbing reports of declining populations of amphibians around the world, especially at high altitudes, and some species are apparently now extinct. This indicates that we may have truly overreached and detrimentally affected the global environment in a profound way.

So even as readers enjoy the fascinating accounts of the living reptiles and amphibians that follow, I hope they will be inspired to act wisely to conserve these creatures in all their marvelous variety.

DR GEORGE B. RABB
Director, Brookfield Zoo;
Vice-Chairman, IUCN Species Survival Commission, Brookfield, USA

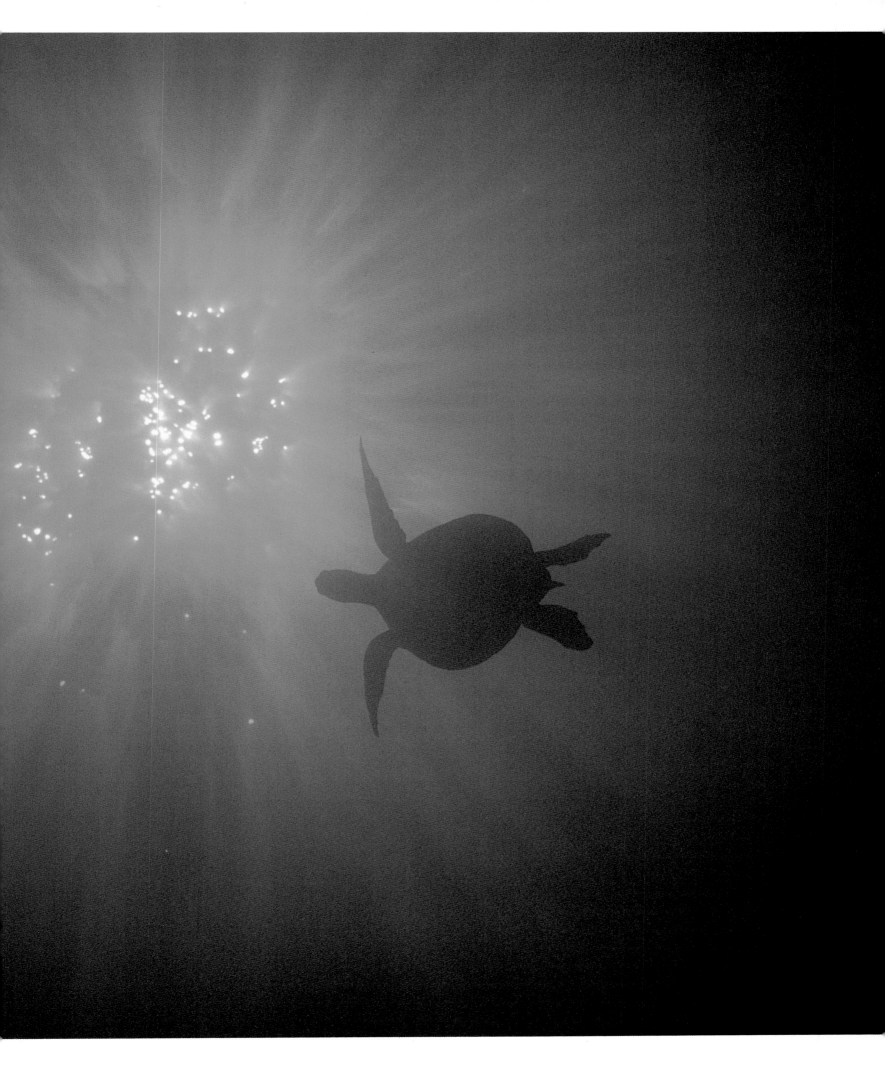

PART ONE

THE WORLD OF
REPTILES & AMPHIBIANS

INTRODUCING REPTILES & AMPHIBIANS

Scientists divide the animal kingdom into several major groups for classification purposes. By far the largest group is the invertebrates: it contains about 95 percent of the millions of known species of animals, including sponges, mollusks, crustaceans, and insects. Reptiles and amphibians are vertebrates—they belong to a group of animals characterized by having a bony backbone, a flexible but strong support column of articulated sections (vertebrae) to which the other body structures are attached. Vertebrates include not only amphibians (class Amphibia) and reptiles (class Reptilia), but also fishes (about three classes), birds (class Aves), and mammals (class Mammalia). The differences between reptiles and amphibians are generally more obvious than their similarities, although their joint study under the name "herpetology" (from the Greek *herpo*, to creep or crawl) is a scientific tradition dating back nearly two centuries.

DIVERSE FORMS

The fossil record shows that both amphibians and reptiles were once much more diverse in size and structure than they are today (dinosaurs being the best known examples), and so there is a tendency to regard today's forms as a mere shadow of their former glory, having been displaced in dominance by the warm-blooded birds and mammals.

However, the amphibians with 4,950 living species, together with 7,400 living reptiles, jointly outnumber the birds (9,000 species) and mammals (about 4,670 species). They are in no way more primitive than birds and mammals. Their behavior and physiology are just as complex and equally well adapted to the astonishingly varied environments they inhabit.

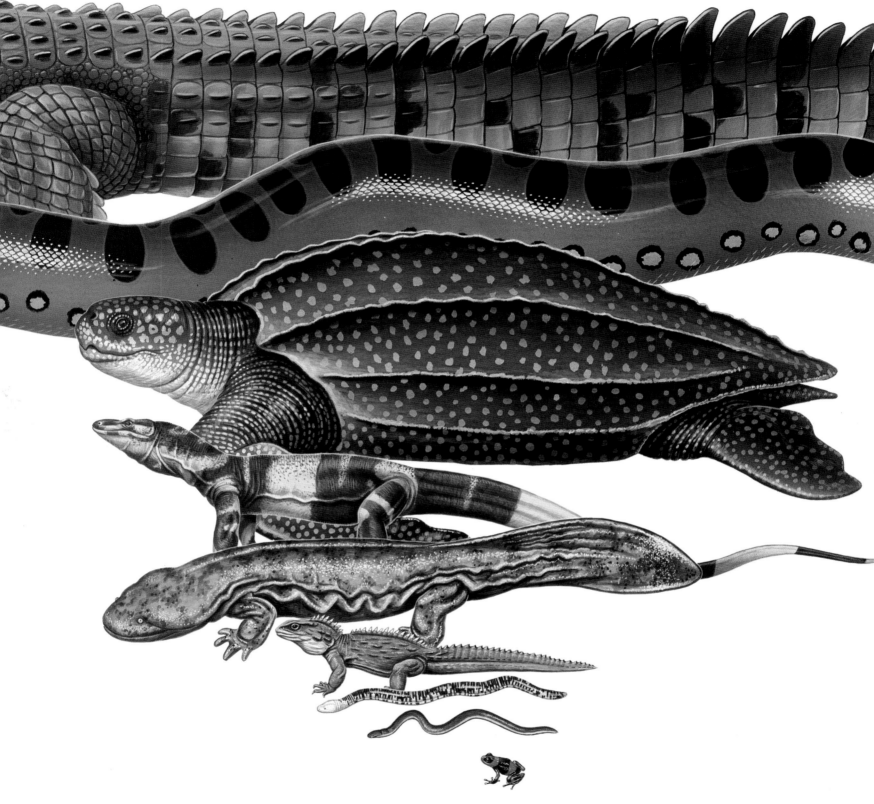

BODY TEMPERATURE AND METABOLISM

One significant factor in the distribution and behavior of modern reptiles and amphibians is their inability to produce sufficient internal metabolic heat to maintain a constant body temperature, as do almost all birds and mammals. For this reason they are often termed "cold-blooded", or "ectotherms". Although some reptiles are able to generate enough metabolic heat to raise their temperature for a specific purpose for limited periods (for example, female pythons brooding their eggs), none can constantly maintain a temperature much higher than that of their surroundings without an external heat source. Their dependence on external sources of heat to reach and maintain an "active" body

▲ Reptiles vary enormously in size and structure, reflecting adaptations suited to almost every conceivable habitat throughout the tropical and temperate regions of the world. Because of their somewhat greater dependence on moisture, amphibians tend to be rather less diverse in size and shape. In both groups, many species are unremarkable in color but others rival birds in their bold patterns and gaudy hues.

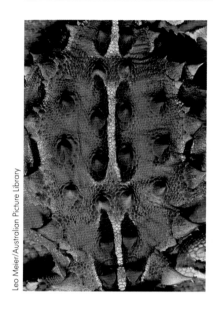

temperature greatly limits their ability to live and breed in colder climates, with the consequence that they are most abundant in tropical and warm temperate regions.

While being cold-blooded can have disadvantages, such as being more vulnerable to predators at low environmental temperatures, it also has a number of advantages. Amphibians and reptiles can simply "shut down" when conditions are unsuitable—for example, if it is too cold or food is scarce—and therefore do not have to use up large amounts of stored energy to keep themselves warm.

SKIN AND SCALES

The skin of most amphibians is thin and highly glandular, and needs to be kept relatively moist to function effectively. It is also highly permeable, allowing water to be absorbed or lost, and is often involved in gas exchange as an adjunct to normal respiration through the lungs. This means that most amphibians are restricted to moist or humid environments.

The skin of reptiles differs from that of amphibians in that the outer layer is thickened to form scales, composed of keratin; some areas are further thickened to form tubercles and crests. In many reptiles (and a few amphibians) small bones develop in the skin just below the

surface. These dermal bones, or osteoderms, add strength and protection to the skin and make it even more impervious to water loss. Skin color and color changes are largely determined by pigment cells just below the outer layers of skin.

BODY STRUCTURE

From the muscular fins of their fish ancestors, the first amphibians evolved four limbs for moving about on land. Subsequent amphibians and reptiles share this basic tetrapod (four-limbed) vertebrate skeletal structure.

This basic skeletal structure has become highly modified in some amphibians and reptiles. An extreme adaptation is found in turtles, in which the greater part of the vertebral column, the ribs, and the limb girdles, have become fused with dermal bones to form a protective shell. In frogs, the development of an efficient jumping mechanism has called for greater rigidity of the vertebral column and a solid structure to support the enormous mass of hind limb muscles. This has been achieved by a great reduction in the number of vertebrae and the incorporation of several vertebrae into a single long supporting bone called the urostyle.

Modern amphibians generally have simple pedicellate teeth, which are used to grasp the invertebrates on which they feed, but in many species the teeth are reduced or absent. Teeth in reptiles vary greatly in form and structure, from simple blunt teeth used to grasp and partially crush prey, to those with broad grinding surfaces or sharp shearing edges. Perhaps the most specialized are the fangs of snakes that have evolved grooves or hollows, making them efficient venom-delivering hypodermic needles.

In most amphibians and reptiles the tongue is an important adjunct to the teeth and jaws in catching prey. It is usually very muscular, flexible, and extrusible, and is often sticky with mucus to adhere to prey such as insects. It is also used to manipulate food within the mouth. In some groups of reptiles, such as monitor lizards and snakes, the tongue has become a specialized sense organ which senses and locates prey, but otherwise is not directly involved in the capturing of prey.

INTERNAL ORGANS

The soft body structures of adult amphibians and reptiles are similar to those of other vertebrate animals. A trachea (windpipe) leads from the glottis to paired lungs, although the lungs may be reduced or absent in some aquatic species that can absorb almost all the oxygen they need through their skin. In most snakes the left lung has been lost during the course of evolution; and in many sea snakes the lung now extends forward along the length of the trachea and is involved in

Leo Meier/Australian Picture Library

▲▼ *(Above) Covered with a water-impermeable barrier of dry, horny scales, like the skin of this Australian thorny devil* Moloch horridus *in close-up, reptiles can in general thrive in much drier habitats than amphibians can tolerate. (Below) In contrast to reptiles, many amphibians can extract some or all of their oxygen requirements through their skin. This is Darwin's frog* Rhinoderma darwinii.

ZEFA/Ziesler

regulating the snake's buoyancy when diving. The circulatory system is complex and efficient, but. with the exception of the crocodiles, the heart of amphibians and reptiles has only three chambers—two auricles or atria and a partially divided ventricle—although the flow is arranged to minimize mixing of oxygenated (arterial) and de-oxygenated (venous) blood in the single ventricle. In crocodiles, as in birds and mammals, the ventricle is completely divided by a septum, which prevents any mixing of arterial and venous blood and is much more efficient than the system found in other reptiles and in amphibians. While many different food specialists have developed among the amphibians and reptiles, once past the mouth the food is digested in a fairly standardized vertebrate digestive system. An important feature of both amphibians and reptiles—in which they differ from most mammals—is the cloaca, a large chamber just inside the vent, into which the gut and the ducts of urinary and sexual organs all enter.

For the majority of amphibians, living as they do in humid environments, water conservation is not a major problem. Consequently their nitrogenous waste products are converted into urea in the liver and then excreted by the kidneys and leave the body in a fluid urine. The same method is used by reptiles that live in moist or aquatic environments, but for reptiles in very dry environments such as deserts, where free water is scarce or rare, body water is too precious to be lost as urine. Consequently in the majority of reptiles the liver converts nitrogenous waste into uric acid which, after excretion by the kidneys, passes to the cloaca where water is resorbed and solid uric acid crystals deposited; these leave the body as a moist paste, using only a fraction of the water that would be needed to excrete urea as urine.

THE SENSES

For most amphibians and reptiles three senses are critical to survival: sight, hearing, and olfaction (smell). The relative importance of each is strongly correlated with the lifestyles of any particular species.

In frogs in which the males attract females by calling to them, hearing clearly needs to be acute and discriminating if the calls of individual males of one species are to be distinguished from the background noise of other calling males or other species. The males of many lizards use brilliant colors to attract mates and to warn other males away from their territories. Clearly, acute vision which recognizes those colors is essential to such species.

The eyes of amphibians and reptiles are similar in general features to those of other land vertebrates, with characteristic cornea, iris, lens, and retina. In most amphibians there is an

immovable upper eyelid and a movable lower eyelid, the upper part of which (the nictitating membrane) is usually transparent. In reptiles the lower eyelid is also usually movable and at least partly scaly, but in many lizards, and in all snakes, the lower and upper eyelid have become fused and the eye is covered by a large, fixed, transparent disk—the spectacle—which protects the eye from damage. The spectacle becomes scratched and dirty over time, and is shed with the skin at each sloughing cycle.

The ears are also typical of other land vertebrates. Sounds are received on a membrane on each side of the head—the tympanum—and transmitted by one or more fine stapedial bones to the inner ear. However, many amphibians and reptiles have limited reception of airborne sounds, as the tympanum is absent or covered by skin, and the inner bony structures are often modified to receive vibrations through other body tissues.

Olfactory senses (smell and taste) are also moderately to well-developed in most amphibians and reptiles. The most important olfactory organ in both amphibians and reptiles is a vomeronasal organ called Jacobson's organ, lying in the roof of the mouth. In amphibians and many reptiles it "tastes" the food in the mouth. In other reptiles, especially monitor lizards and snakes, the protruding tongue "tests" its environment by

▲ *Mammals and birds shed and replace their skins constantly, cell by cell, but reptiles typically do so all at once, like this sloughing Bibron's gecko* Pachydactylus bibroni *of Africa.*

▼ *Containing both rods and cones, the eye of a crocodile is thought to be capable of color vision.*

David P. Maitland/AUSCAPE International

collecting minute particles which are passed for identification to Jacobson's organ by the tip(s) of the tongue. In this way such reptiles can follow the trail even of moving prey.

REPRODUCTION AND LIFE CYCLES

The life cycle of an amphibian is fundamentally different from that of a reptile. Reptiles produce eggs within which the embryo develops and hatches (or is born, in the case of live-bearing species), generally as a miniature replica of the adult. Amphibians are characterized by a two-stage life cycle in which the eggs hatch into aquatic, gill-bearing larvae (for example, tadpoles) which eventually metamorphose into air-breathing, mostly terrestrial adults. Occasionally the entire tadpole stage takes place within the egg or within the body of the mother.

Almost all amphibians produce eggs, varying in number from a single egg to many thousands in a clutch, which are deposited in water or in humid sites on land. They are fertilized externally or internally. Most male amphibians lack any form of intromittant organ with which to introduce sperm inside the female, but some are able to protrude part of their cloaca into that of the female. In most frogs, the male simply expels his seminal fluid over the eggs as they are extruded from the female's body. In salamanders, the male deposits a package of sperm (the spermatophore), which is picked up by the cloacal lips of the female and is held in her cloaca until needed for fertilization.

The egg of a reptile is fertilized internally. With the exception of the tuatara, in which sperm are exchanged through cloacal contact, reptiles have well-developed intromittant organs to deliver the seminal fluid. In lizards and snakes these organs are paired, so that each male has two functioning penises. Turtles and crocodilians have only a single penis.

The reptile egg is substantially more complex than that of an amphibian, and much better adapted to survival on land. In many reptiles, the egg does not develop a shell, and development occurs within the mother's body; the embryonic membranes are retained and in some cases form a primitive placental connection with the mother.

HAROLD G. COGGER

REPTILE & AMPHIBIAN EGGS

The eggs of amphibians do not have a shell or any other protective membrane and are therefore very vulnerable to desiccation. Each fertilized egg is surrounded by a dense protective jelly, and contains yolk to nourish the developing young until it hatches; the embryo's waste products simply permeate out through the surrounding jelly.

Most modern reptiles lay eggs, but they differ significantly from those of amphibians. Typically, the developing embryo is cushioned by a series of fluid-filled sacs within a protective calcareous shell. The shell physically protects the embryo and although permeable to gases, and somewhat permeable to water, greatly reduces water loss, allowing the egg to develop safely in relatively dry locations on land. Inside the shell the yolk provides food for the developing embryo, and three embryonic membranes carry out special functions. The allantois is a sac in which the embryo's waste products are safely stored. It is partly fused with another sac, the chorion, which lies against the inner surface of the shell, and both contain blood vessels, which assist in exchanging oxygen needed by the embryo through the permeable egg shell. The third membrane, the amnion, forms a fluid-filled sac around the embryo, which cushions it from external disturbance and prevents dehydration.

This kind of egg, unlike that of an amphibian, is essentially a closed system, in which all of the embryo's basic needs are met. It is known as a cleidoic egg.

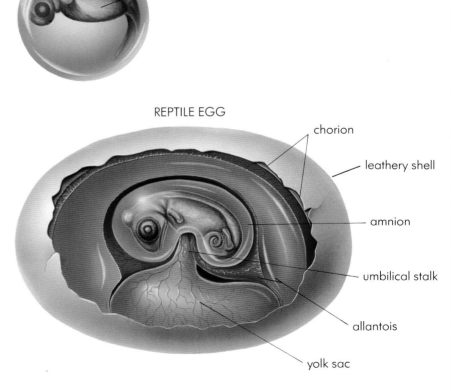

AMPHIBIAN EGG

— yolk

REPTILE EGG

chorion

leathery shell

amnion

umbilical stalk

allantois

yolk sac

▲ A typical amphibian egg (top) and a reptilian egg, showing membranes and leathery shell (bottom).

CLASSIFYING REPTILES & AMPHIBIANS

Biologists have discovered, described, and given names to about 1.5 million of the many millions of species of plants, animals, and microorganisms that exist at present. They have also named many species that formerly lived on Earth but are now extinct. The role of classification is to provide a unique name for each of these organisms and put them in a hierarchy of increasingly inclusive taxa (groups) based on their evolutionary relationships, so that they can be unequivocally recognized and associated with other species having a common ancestry.

▼ A red-eyed leaf-frog Agalychnis callidryas clasps a heliconia in a Costa Rican rainforest. One obvious external feature sets frogs and toads apart from all other amphibians: their hind limb consists of four hinged elements, not three.

SYSTEMATICS AND TAXONOMY

The science of discovering the diversity and evolutionary relationships of organisms is called systematics, and aspects of systematics involved with naming and classifying organisms form the subdiscipline of taxonomy. The basic taxon (group) is the species, and ideally all higher or more inclusive taxa comprise an ancestral species and all its descendants. Determination of this relationship is based on all members of a group sharing one or more derived features (evolutionary novelties). For example, all frogs and toads have two of the ankle bones greatly elongated to form a fourth segment in the hind leg (all other limbed amphibians and reptiles have three segments in the hind leg). This condition provides greater leverage for leaping and serves to distinguish the group, the order Anura, in which frogs and toads are placed.

NAMES AND CLASSIFICATION

Scientists use Latinized names for groups of organisms in order to provide universality and stability and to avoid the necessity of translating the name into many different languages. No matter what one's native language, the Latinized scientific name is immediately associated with the same group of organisms.

The fundamental category in all classifications is the species. Species are populations of organisms that share one or more similarities not found in related species. Species form closed · genetic systems, although some closely related species may occasionally hybridize. The name of a species is composed of two words: the first is the name of its genus (the generic name) and indicates its closest relationships; the second name (the specific name) is uniquely used for that species within a genus. The generic name always

J. Cancalosi/AUSCAPE International

▲ Even such intangible areas as vocalizations and behavior can yield features of value in classifying animals, and in principle there is no reason why any one set of features is inherently any more valid than another. But for obvious, practical reasons those features involving external appearance and structure (morphology) are most useful in the widest range of circumstances. For example, careful inspection of the scales on a snake's head, like this carpet python Morelia spilotas, reveals patterns that vary from one group to another, and are useful in distinguishing them.

the nerve cord and extends the length of the body; and gill arches and gill pouches (which are present only embryonically in reptiles, birds, and mammals). In other chordates, including larval amphibians, the gill arches support functional gills and the gill pouches in the throat region open to the outside.

Three subphyla are recognized within the Chordata; amphibians and reptiles are located in the subphylum Craniata. Animals belonging to this group have chambered hearts, semicircular canals in the inner ear, skeletal gill supports, and a protective cartilaginous or bony cranium around the brain. Most members of this group also have segmentally arranged skeletal elements (vertebrae) surrounding and protecting the spinal cord.

The next division in the hierarchy is the class. Amphibians and reptiles are placed in different classes, the Amphibia and Reptilia, respectively. Classes are further divided into orders; there are three living orders of amphibians and four of reptiles. Orders are divided into families, families into genera, and each genus comprises one or more species. The classification adopted in this book also makes use of the suborder, which is a category more inclusive than a family but less inclusive than an order.

These hierarchical categories form a set of taxa defined by shared similarities. This means that species within a genus are more similar to one another than to species in other genera within the same family; genera in the same family are more similar to one another than to genera in other families; and so on. It also means that, progressing up the hierarchy, fewer and fewer features are shared by the species included in an order, a class, a phylum, or a kingdom. The hierarchy thus reflects in reverse the branching pattern of the evolution of life, because the decrease in similarity correlates with time of divergence—that is, the greater the similarity the more recent the divergence.

begins with a capital letter, and the specific name is always written in the lower case, and both names are always printed in *italics*; for example, the name of the Gaboon viper of tropical Africa is written *Bitis gabonica*. Species names are frequently based on a distinctive feature, or may refer to the species' place of origin, or may honor a person by using a Latinized form of their name.

In some instances when populations of any one species are separated geographically and have relatively consistent differences, they may be recognized as subspecies and then have a third name: for example, *Lampropeltis triangulum triangulum* for the eastern milk snake and *Lampropeltis triangulum gentilis* for the plains milk snake of North America.

Scientists also use a standard set of terms for each level in the hierarchical arrangement of taxa and a single Latin name (always capitalized) for each different taxon. The most inclusive category to which amphibians and reptiles belong is the kingdom Animalia. This taxon contains all organisms that are multicellular, obtain their energy by ingesting food from the surroundings, and usually develop from the coming together of two cells, a large egg and a small sperm.

The kingdom Animalia is composed of 35 phyla, each phylum representing a distinctive major biological organization and lifestyle. Amphibians and reptiles belong to the phylum Chordata, which consists of animals that have a single dorsal nerve cord; a cartilaginous notochord (at least embryonically) that lies below

VERNACULAR AND COINED NAMES

Amphibians and reptiles are known by vernacular or common names that vary from language to language, country to country, and even within the same country. In some cases, the same name may be used for a number of similar appearing species. In others, one name is used for entirely different creatures; for example, in different parts of the United States the name "gopher" is applied to a turtle, a ground squirrel, and a subterranean rodent called a pocket gopher. Because of the potential for confusion these names are not used in formal, scientific identification. Species names that had never been used by local people for amphibians and reptiles have in recent years been coined for use in popular works or field guides, especially in English. These names, alas, lack even the dignity of being truly vernacular, as

a

they are inventions of their authors, duplications of scientific names, and usually uninformative when read by someone who has a different native language.

SPECIES

There are about 4,950 species of living amphibians and about 7,400 species of living reptiles. Although some of the larger and more spectacular representatives—such as the "hairy frog", the leatherback turtle, and the Komodo dragon—are very distinctive, many closely related forms are characterized by subtle differences. External form and structure, size and body proportions, and coloration are usually the basis for separating allied species, with differences in the number, shape, and structure of scales being emphasized in reptiles. Recent studies of chromosomes have been valuable in distinguishing between cryptic species—those whose external features differ only slightly from their closest relatives.

Because many kinds of amphibians and reptiles are small and secretive, occur very locally, or are inhabitants of unexplored regions of the Earth (especially tropical rainforests), previously undescribed species are discovered each year. During the past decade about 80 new species of amphibians (mostly frogs and toads) and 25 species of reptiles (mostly lizards) have been described each year. Sometimes, as our knowledge of a particular group increases, forms previously thought to be distinctive are shown to belong to a single species. These factors, plus the cases in which conflicting evidence (or its interpretation) leads different scientists to disagree on the status of named forms, reflect the differing numbers of species recognized by one or another authority. It is safe to say, however, that new species of amphibians and reptiles will continue to be discovered at these rates well into the future. It should be noted that while amphibians and reptiles are often regarded as minor remnants of an earlier evolutionary radiation, the number of species of living amphibians exceeds that of mammals, and there are many more species of living reptiles than of mammals.

GENERA

A genus contains one or more species that share at least one derived, or advanced, feature. The classification of amphibians into different genera is usually based on internal features, especially those of the skeleton, because their smooth bodies possess relatively few significant external differences. Skeletal morphology is also used in reptiles, but scalation (especially the scales on the head) frequently provides a basis for generic groupings. Occasionally a single, very distinctive living species may be classified in a genus of its own because of one or more unique features; for example, most American and Eurasian treefrogs are placed in the genus *Hyla*, but the spiny-headed or crowned treefrog is placed in a separate genus, *Anotheca*, as *Anotheca spinosa*. Frequently, however, genera originally thought to contain only a single species have newly discovered living or fossil species added to them as the result of additional research. Currently we recognize about 430 genera of living amphibians and 898 genera of living reptiles.

FAMILIES

Genera are clustered into different families on the basis of one or more shared features of internal morphology. Frequently it is difficult for the layperson to readily identify which family a species belongs to, from only its external features; even a specialist may have a problem with family placement of unfamiliar amphibians, for external resemblances are deceiving. A few families of amphibians have only a single very distinctive genus containing one or a few species. In theory such families should be regarded as sole relics of a larger group of species, the majority of which have become extinct.

All family names end in the suffix "-idae", and because many groups of amphibians and reptiles lack vernacular names the family name is often rendered into English as a word ending in "-id"; for example, the lizard family Agamidae may be called agamids. Families may be divided into two or more subfamilies when there are marked differences among the included genera. Amphibians are placed into 43 families represented by living species, and reptiles are placed into 64 families.

▼ *A species is essentially a closed genetic system: interbreeding takes place between individuals within the system but not outside it. Usually such systems have close neighbors, forming clusters called genera. But in some cases most of the species forming the cluster die out, leaving perhaps only a single survivor at the present time. This probably happened in the case of the crowned frog Anotheca spinosa of Central America, a species that has no close relatives anywhere.*

Michael Fogden/Oxford Scientific Films

Tom Ulrich/Oxford Scientific Films

▲ *Now confined to a few tiny islands in New Zealand, the two species of tuatara are the sole surviving remnants of an entire order of reptiles that had their heyday some 200 million years ago, but died out one by one during the intervening eons.*

ORDERS

Orders are clusters of families that share one or more derived features and are obviously separated from other such groups. Most species of amphibians may be immediately recognizable as a salamander or newt (order Caudata), a limbless caecilian (order Gymnophiona), or a long-limbed, tailless frog or toad (order Anura). Similarly, it is difficult to confuse a turtle (order Testudinata), a typical lizard, amphisbaenian, or snake (order Squamata), or a crocodilian (order Crocodilia)

with one another. The tuatara of New Zealand superficially resemble lizards but are unlike any lizard in obvious details of scalation and numerous internal features. Tuatara are the sole living representatives of the order Rhynchocephalia, but a number of extinct rhynchocephalians are known from the Mesozoic, 245 to 65 million years ago.

There are also many extinct orders of amphibians and reptiles with no representatives alive today.

HIGHER CLASSIFICATION

Scientists agree that modern amphibians are related to one another and in turn are related to a cluster of ancient amphibians that date back to the upper Devonian, about 380 million years ago, but became extinct by the end of the Triassic, 200 million years ago. The difficulty of interpreting these relationships and the status of other fossil "amphibians"—including the lineage most closely allied to early reptiles—leaves the higher classification of all so-called "amphibians" in a state of flux. At this time, it seems best not to use the conflicting classification schemes, except to note that the three living orders are usually regarded as forming the subclass Lissamphibia.

Reptiles, on the other hand, form two distinct subclasses based primarily on differences in skull and jaw characteristics. The subclass Parareptilia includes several extinct lines. The living reptiles and many extinct groups are now recognized as part of the subclass Eureptilia and its three

THE WARTY NEWT: AN EXAMPLE OF CLASSIFICATION

The concepts of zoological classification and nomenclature may be illustrated by using the warty newt *Triturus cristatus* of western Eurasia as an example. The genus *Triturus* contains 12 species of newts all from this region. *Triturus* is grouped with a series of 14 other genera found in the temperate zone of North America and eastern Eurasia and the temperate and subtropical zones of eastern Asia, into the family Salamandridae. Together with a number of other families of amphibians, in which the adults have limbs and a tail and their aquatic larvae have true teeth, the Salamandridae belong to the order Caudata. This order in turn belongs to the class Amphibia. The Amphibia are included with the class Reptilia, class Mammalia and several classes of aquatic jawed and jawless fishes, in the subphylum Craniata which, together with tunicates (subphylum Urochordata) and lancelets (subphylum Cephalochordata) constitute the phylum Chordata. The Chordata are one of 35 phyla of multicellular, heterotrophic organisms that form the kingdom Animalia.

Stephen Dalton/NHPA

▲ *The warty, or crested newt Triturus cristatus of Europe.*

Kingdom: Animalia
Phylum: Chordata
Subphylum: Craniata
Class: Amphibia

Order: Caudata
Family: Salamandridae
Genus: *Triturus*
Species: *Triturus cristatus*

superorders. The superorder Lepidosauria contains many fossil forms (such as ichthyosaurs and plesiosaurs) plus the tuataras, lizards, and snakes. The superorder Testudines comprises turtles and the fossil saurpterygians. The superorder Archosauria includes the extinct dinosaurs, pterosaurs, and phytosaurs, the living crocodilians, and technically the birds, although the birds are traditionally placed in a separate class.

CHANGES IN CLASSIFICATION

Classifications are modified to reflect newly understood evolutionary relationships. The recent integration of chromosomal and molecular data into evolutionary studies, and better ways of evaluating existing classifications, demonstrate this process. Classifications evolve and they must be understood as impermanent statements of current knowledge. The present framework that recognizes three orders of living amphibians and four orders of living reptiles seems firmly established, but some revision in the number and positioning of families, especially for those with larger numbers of species, is to be expected as relationships become clearer.

JAY M. SAVAGE

LIVING ORDERS, SUBORDERS, AND FAMILIES OF AMPHIBIANS AND REPTILES

Because the classification of amphibians and reptiles is undergoing constant revision, there is no definitive classification list available. The following represents a consensus of the academics who have contributed to this publication. In the case of several large groups, additional taxonomic categories between suborder and family (infraorders and superfamilies) have been indicated to show relationships.

CLASS AMPHIBIA

SUBCLASS LISSAMPHIBIA

ORDER
CAUDATA — **SALAMANDERS & NEWTS**

Suborder Sirenoidea
Sirenidae — Sirens

Suborder Cryptobranchoidea
Cryptobranchidae — Hellbenders & giant salamanders
Hynobiidae — Hynobiids

Suborder Salamandroidea
Amphiumidae — Amphiumas (Congo eels)
Plethodontidae — Lungless salamanders
Rhyacotritonidae — Torrent salamanders
Proteidae — Mudpuppies, waterdogs, & the olm
Salamandridae — Salamandrids
Ambystomatidae — Mole salamanders
Dicamptodontidae — Dicamptodontids

ORDER
GYMNOPHIONA — **CAECILIANS**
Rhinatrematidae — South American tailed caecilians
Ichthyophiidae — Ichthyophiids
Uraeotyphlidae — Uraeotyphlids
Scolecomorphidae — Scolecomorphs
Caeciliidae — Caeciliids & aquatic caecilians

ORDER
ANURA — **FROGS & TOADS**
Ascaphidae — "Tailed" frogs
Leiopelmatidae — New Zealand frogs
Bombinatoridae — Fire-bellied toads & allies
Discoglossidae — Discoglossid frogs
Pipidae — Pipas & "clawed" frogs
Rhinophrynidae — Cone-nosed frog
Megophryidae — Megophryids
Pelodytidae — Parsley frogs
Pelobatidae — Spadefoots
Superfamily Bufonoidea
Allophrynidae — Allophrynid frog
Brachycephalidae — Saddleback frogs
Bufonidae — Toads
Helophrynidae — Ghost frogs
Leptodactylidae — Neotropical frogs
Myobatrachidae — Australasian frogs
Sooglossidae — Seychelles frogs
Rhinodermatidae — Darwin's frogs
Hylidae — Hylid treefrogs
Pelodryadidae — Australasian treefrogs
Centrolenidae — Glass frogs

Pseudidae — Natator frogs
Dendrobatidae — Dart-poison frogs
Superfamily Ranoidea
Microhylidae — Microhylids
Hemisotidae — Shovel-nosed frogs
Arthroleptidae — Squeakers
Ranidae — Ranid frogs
Hyperoliidae — Reed & lily frogs
Rhacophoridae — Rhacaphorid treefrogs

CLASS REPTILIA

SUBCLASS EUREPTILIA

SUPERORDER LEPIDOSAURIA
ORDER
RHYNCHOCEPHALIA — **TUATARAS**
Sphenodontidae — Tuataras

ORDER
SQUAMATA — **SQUAMATES**
Suborder Iguania — **IGUANID LIZARDS**
Corytophanidae — Helmeted lizards
Crotaphytidae — Collared & leopard lizards
Hoplocercidae — Hoplocercids
Iguanidae — Iguanas
Opluridae — Madagascar iguanians
Phrynosomatidae — Scaly, sand & horned lizards
Polychrotidae — Anoloid lizards
Tropiduridae — Tropidurids
Agamidae — Agamid lizards
Chameleonidae — Chameleons

Suborder Scleroglossa
Superfamily Gekkonoidea
Eublepharidae — Eye-lash geckos
Gekkonidae — Geckos
Pygopodidae — Australasian flapfoots
Superfamily Scincoidea
Xantusiidae — Night lizards
Lacertidae — Lacertids
Scincidae — Skinks
Dibamidae — Dibamids
Cordylidae — Girdle-tailed lizards
Gerrhosauridae — Plated lizards
Teiidae — Macroteiids
Gymnophthalmidae — Microteiids
Superfamily Anguioidea
Xenosauridae — Knob-scaled lizards
Anguidae — Anguids; glass & alligator lizards
Helodermatidae — Beaded lizards
Varanidae — Monitor lizards
Lanthanotidae — Earless monitor lizard

Suborder Amphisbaenia — **AMPHISBAENIANS**
Bipedidae — Ajolotes
Amphisbaenidae — Worm lizards
Trogonophidae — Desert ringed lizards

Rhineuridae — Florida worm lizard

Suborder Serpentes — **SNAKES**
Infraorder Scolecophidia
Anomalepididae — Blind wormsnakes
Typhlopidae — Blind snakes
Leptotyphlopidae — Thread snakes
Infraorder Alethinophidia
Anomochelidae — Stump heads
Aniliidae — Coral pipesnakes
Cylindrophidae — Asian pipesnakes
Uropeltidae — Shield tails
Xenopeltidae — Sunbeam snake
Loxocemidae — Dwarf boa
Boidae — Pythons & boas
Ungaliophiidae — Ungaliophiids
Bolyeriidae — Round Island snakes
Tropidophiidae — Woodsnakes
Acrochordidae — File snakes
Atractaspididae — Mole vipers
Colubridae — Harmless & rear-fanged snakes
Elapidae — Cobras, kraits, coral snakes & sea snakes
Viperidae — Adders & vipers

SUPERORDER TESTUDINES
ORDER
TESTUDINATA — **TURTLES, TERRAPINS & TORTOISES**
Suborder Pleurodira — **SIDE-NECK TURTLES**
Chelidae — Snake-neck turtles
Pelomedusidae — Helmeted side neck turtles

Suborder Cryptodira — **HIDDEN-NECKED TURTLES**
Superfamily Trionychoidea
Kinosternidae — Mud & musk turtles
Dermatemydidae — Mesoamerican river turtle
Carretochelyidae — Australian softshell turtle
Trionychidae — Holarctic & Paleotropical softshell turtles
Superfamily Chelonioidea
Dermochelyidae — Leatherback sea turtles
Cheloniidae — Sea turtles
Superfamily Testudinoidea
Chelydridae — Snapping turtles
Emydidae — New World pond turtles & terrapins
Testudinidae — Tortoises
Bataguridae — Old World pond turtles

SUPERORDER ARCHOSAURIA
ORDER
CROCODILIA — **CROCODILIANS**
Alligatoridae — Alligators & caimans
Crocodylidae — Crocodiles
Gavialidae — Gharials

REPTILES & AMPHIBIANS THROUGH THE AGES

The first backboned animals (vertebrates) evolved in the sea at the end of the Cambrian period about 500 million years ago. During the next 150 million years they continued to evolve in the world's oceans into amazingly diverse groups of fishes. But to colonize the land these aquatic vertebrates had to meet and overcome the tremendous challenges involved in exchanging an aquatic existence for a terrestrial one. These included breathing atmospheric rather than dissolved oxygen, abandoning the buoyancy of water, and modifying the body structure to meet the high gravitational forces encountered on land.

FROM WATER TO LAND

The first vertebrates to make the transition from water to land were the amphibians. In the late Devonian period about 360 million years ago, they became the first tetrapods—that is, the first backboned animals with four articulated legs—from which all later vertebrates (reptiles, birds, and mammals) evolved. Indeed, once the amphibians had evolved the basic structures needed to colonize the land, the evolutionary opportunities for vertebrates were immense. Only about 50 million years after the first amphibians appeared, a relatively short time in geological terms, the first reptiles evolved from an amphibian ancestor, and by the Triassic period about 240 million years ago they had become the dominant land-dwelling animals. This was the beginning of the era of the mighty dinosaurs.

FROM FISHES TO AMPHIBIANS

The origin of early tetrapods is still a matter of scientific debate. Among the animals alive today, the closest relatives to the tetrapod ancestor are probably the lungfishes. They too have internal nostril openings (choanae), allowing air to be taken in while the mouth is closed, as well as lungs and a functionally divided heart.

▼ Eusthenopteron was a lobe-finned fish of the Devonian period, close to, but not directly on, the evolutionary line that led to the amphibians. It had lungs as well as gills and could breathe air during excursions over land when its pond dried up. Its paired fins, supported by muscular lobes with bony elements, were the precursors of the tetrapod limb bones.

Richard Hammond UK (model); Carl Bento, Australian Museum (photography)

Another living fish sharing similarities with tetrapods is the famous coelacanth, which was discovered in the seas near South Africa as recently as 1938, bearing testimony to the survival of a group thought to have died out 65 million years ago. The coelacanth's paired fins make the kind of movements that we assume tetrapod ancestors made, although its anatomical details are not closely related to those of tetrapods. Fossils of lobe-finned fishes of the Devonian period show closer similarities, both in skull structure and in the structure of the pectoral fins.

EARLY AMPHIBIANS

The earliest known amphibians, *Ichthyostega* and *Acanthostega,* whose fossils were found in Greenland, lived about 360 million years ago. They belong to a group known as labyrinthodonts, named for the labyrinthine infoldings of the pulp cavity of their teeth. These early forms are interesting because they retain some fish-like characteristics, such as a fish-like tail and a rudiment of the bony gill cover in the cheek region of the skull. In fish the bone supporting the bony gill cover also supports the jaw articulation. Later, during the course of tetrapod evolution, this bone was eventually reduced to a slender rod (the stapes) capable of transmitting sound waves picked up by a tympanic membrane (the eardrum) to the pressure-sensitive inner ear.

In fish the pectoral or shoulder girdle forms the back end of the skull. In tetrapods such as amphibians and reptiles, this girdle is not attached to the skull but is separated from it by a neck which allows the skull to move more freely. The pectoral and pelvic girdles anchor and support the limbs, and the vertebral column becomes suspended between the pectoral and pelvic girdles like a bridge, requiring not only modification of soft tissues but also the evolution

of special joints between the vertebrae to permit the backbone to bend sideways. Simultaneous breathing and locomotion may not have been possible for the early amphibians if they ventured on land, because the rib movements needed to ventilate their lungs may have conflicted with the muscle contraction required to undulate the vertebral column from side to side.

"REPTILIOMORPH" AMPHIBIANS

One very primitive tetrapod from the early Carboniferous period about 340 million years ago was *Crassigyrinus,* whose fossils were found in Scotland. From its rather weak limbs, we assume it had aquatic habits. It too retained a rudiment of the original gill cover and had skull proportions similar to those of lobe-finned fishes. But the structure of the skull indicates that *Crassigyrinus* may be related to the Anthracosauria, one of the major divisions of labyrinthodont amphibians.

The anthracosaurs are now considered to belong to the "reptiliomorph" group, including the ancestor of reptiles, birds, and mammals—as opposed to the "batrachomorph" group, which included the ancestor of modern amphibians.

The reptiliomorphs were predominantly aquatic amphibians and retained traces of a lateral line canal system in the skull bones (used to detect changes in water pressure or water currents) and had an elongated body with short limbs adapted for eel-like locomotion. A few had well-developed limbs, such as *Proterogyrinus,* a predominantly terrestrial animal up to 60 centimeters (24 inches) long.

During the Permian period 285 to 245 million years ago, the evolutionary trend in some amphibians seems to have been toward adaptations for terrestrial life. Fossils have been found in "red beds" (red sandstone) in locations that in the Permian were dry coastal plains. *Seymouria,* from the Permian red beds of Texas,

▲ Acanthostega, *one of the earliest known amphibians, had a skull similar to that of its fish ancestors, but also had strong limbs, indicating its shallow-water or even terrestrial habits. However, it probably never ventured far from water because it retained a bone-supported fin on its tail and lateral line sensory canals in its skull.*

was about 60 centimeters (24 inches) long, rather massively built, and would have lived mainly on land; it was a reptiliomorph in a number of features, most notably the skull and the vertebral column. *Diadectes*, the earliest tetrapod that can be positively identified as a plant-eater, was up to 3 meters (10 feet) long. The fact that this reptiliomorph amphibian was for a while classified as a reptile highlights the difficulty scientists have in separating reptiles from amphibians on their skeletons alone.

"BATRACHOMORPH" AMPHIBIANS
Among the earliest labyrinthodonts were two representatives of the "batrachomorph" amphibians (Greek *batrakhos* means "frog"), which lived during the early Carboniferous period about 360 million years ago. They are classified in the order Temnospondyli. *Caerorhachis* was predominantly terrestrial, and unlike the reptiliomorphs (and all modern reptiles and amphibians) it did not have the "otic notch", a concavity in the skull above the cheek region which accommodates the tympanic membrane. *Greererpeton*, an aquatic animal, also had no sign of this otic notch; in its skull the stapes was a massive element (similar to that in fossils of lobe-finned fishes) supporting the braincase. The development of a middle ear structure occurred only later in batrachomorph evolution, among the Temnospondyli, apparently independent from the reptiliomorph group. This suggests that the middle ear of modern amphibians has quite a different origin from the middle ear of reptiles and mammals.

Dendrerpeton from the late Carboniferous period of North America, about 290 million years ago, was one of the earliest representatives of the Temnospondyli with an otic notch and predominantly terrestrial habits. In later representatives of this group, such as the larger, massively built *Eryops* and the heavily armored *Cacops*, both found in North America, the skull was still relatively high and narrow, with the eyes on the sides but quite high; the vertebrae and the ribcage were well developed; and strong girdles and limbs supported the animals on their predatory excursions. During the Triassic period when seas inundated the northern continents, many amphibians grew to astonishing size, the body was elongated, and the limbs reduced; in *Metoposaurus* and *Cyclotosaurus* the skull was large (up to 1 meter, or 3¼ feet long) and flat, with the eyes on the top. Among other representatives of the Temnospondyli were the Branchiosauridae, a diverse family whose fossils are abundant in Europe, including not only the aquatic larvae of predominantly terrestrial adults, but also creatures that reached sexual maturity as an aquatic larva, like the axolotl today (order Caudata).

It is generally accepted, on the basis of a number of characteristics (most importantly the structure of the inner ear and the pedicellate teeth), that all modern amphibians have a common ancestral origin within Temnospondyli.

MODERN AMPHIBIANS
The class Amphibia, subclass Lissamphibia, is represented by the three living orders of amphibians: the Anura (frogs and toads), the Caudata (salamanders and newts), and the Gymnophiona (caecilians).

Perhaps the ancestors of today's amphibians belonged to one particular branch of the Temnospondyli. *Doleserpeton* was a small terrestrial amphibian of the early Permian period about 285 million years ago. Fossil bones found in Oklahoma, United States, share with modern amphibians many common features in vertebral structure. *Doleserpeton* also had teeth of a type known as pedicellate, which occur in today's amphibians, in which the tooth is divided horizontally by unmineralized tissue—quite unlike the labyrinthodont infolding described earlier. Most fossil skulls of *Doleserpeton* are smaller than 12 millimeters (½ inch) in length, so it is possible that they are the skulls of immature animals, and the peculiar tooth structure may be a juvenile feature.

The fossil record of lissamphibians is rather poor, frogs being the notable exception. *Triadobatrachus* from the early Triassic of Madagascar, 245 million years ago, resembled modern frogs in a number of features such as the skull and the elongated hind limbs.

The earliest "true" salamander fossil was found in Russia and lived during the late Jurassic, about 150 million years ago. The earliest caecilian fossil is from the late Cretaceous, more than 65 million years ago, but only a vertebra was found. However, evidence suggests that the caecilian group is as old as the break-up of the Gondwana landmass into the southern continents, which occurred some 200 million years ago.

THE FIRST REPTILES
The oldest supposed reptile fossil comes from the early Carboniferous of Scotland and Nova Scotia about 340 million years ago. By 315 million years ago the Captorhinomorpha had evolved; this was a group of rather small and agile lizard-like animals known from fossils found in tree stumps of the giant lycopod *Sigillaria* tree of the late Carboniferous, as well as in subsequent Permian sediments. Skull proportions had changed, so that these predators no longer had the powerful snapping bite characteristic of labyrinthodont amphibians, but had a hard crushing bite suited to deal with the chitinous armor of their insect prey. The Captorhinomorpha represent a basal group of reptiles from which it is believed all other reptiles evolved.

Until recently, classifications of reptilian diversity have been based mainly on the structure of the skull, with up to five subclasses being recognized. The oldest of these subclasses, which contained the captorhinomorphs and so gave rise to all other reptiles, is the Anapsida. In this group the bony covering of the upper cheek region of the skull is complete, although in turtles (the only surviving anapsids) this region of the skull may have openings around the edges to accommodate various muscles. Their fossil record reaches back into the late Triassic with *Proganochelys*, which differed from modern turtles only in having teeth on the palate and a few other features. Turtle diversity increased in the Jurassic, but has subsequently declined. Despite their apparent ponderousness, turtles have survived on Earth with surprisingly little change in their basic structure for nearly 200 million years.

The Synapsida is a group that appeared in the late Carboniferous, about 300 million years ago, and is characterized by a single low opening in the cheek region of the skull. The earliest synapsids were the pelycosaurs, a diverse group which contained both swift-moving carnivorous reptiles and large, heavily-built plant-eating reptiles. Evolving from pelycosaurs in the mid-Permian period about 250 million years ago, and surviving for nearly 50 million years, were the therapsid or mammal-like reptiles. Formerly believed to have descended from captorhinomorph ancestors, the Synapsida and their descendants, the mammals, are now recognized as a separate evolutionary lineage.

The Diapsida is a group characterized by one (upper) or two (upper and lower) openings in the cheek region of the skull. The earliest example is *Petrolacosaurus* from the late Carboniferous of Kansas. The Diapsida split into two major lineages, the Archosauromorpha (the "ruling reptiles"), and the Lepidosauromorpha, which include the squamates (lizards, amphisbaenians, and snakes) and the tuatara.

The Archosauromorpha is the most diverse group of reptilian evolution. Early representatives were the plant-eating rhynchosaurs which, together with mammal-like reptiles (synapsids), were the dominant animals on land for perhaps 60 million years or more—until the rise of those best known archosaurs, the dinosaurs and their relatives, with their improved locomotor abilities and perhaps more sophisticated temperature-regulating physiology. Along coastal areas lived the prolacertiforms, some highly agile, predominantly terrestrial predators of about 80 centimeters (31 inches) total length, such as *Macrocnemus*. Perhaps the oddest of all was *Tanystropheus*, which lived in the sea preying on fish and cephalopods (squid, octopus, etc.), reaching out at them with its grotesquely elongated neck, as long as body and tail together. Its total length was 6 meters (almost 20 feet).

THE AGE OF DINOSAURS

Dinosaurs ruled the Earth for 140 million years, throughout the Jurassic and Cretaceous periods. This reign came to an abrupt end when, some 65 million years ago, a catastrophic event wiped out the dinosaurs and all other large reptiles, except the crocodiles and turtles.

The early dinosaurs split into two groups: the Saurischia (with a reptile-like pelvis), and the

▼ The discovery of Deinonychus, a saurischian, helped to shed the myth of the clumsy, sluggish, tail-dragging dinosaur. Grabbing its prey with long forelimbs, this active predator disembowelled its prey with scythe-like claws on its hind limbs, a feat requiring great speed and agility. Its tail was stiffened with bony rods, enabling it to be held well off the ground for balance.

European Alps, evolved a much-enlarged fourth digit which supported a wing membrane. The rhamphorhynchoids of the subsequent Jurassic period remained rather small and may superficially have resembled modern bats, but the pterodactyloids of the Cretaceous period included the largest animals that ever took to active flight. The best known is *Pteranodon* with a wingspan of 7.5 meters (24 ½·feet), and there is evidence that related forms such as *Quetzalcoatlus* had a wingspan of 11 to 12 meters (36 to 39 feet).

But even the pterosaurs were not the first vertebrates invading the air. Perhaps 100 million years earlier, during the late Permian, there was a small group of lizard-like diapsid reptiles, the coelurosauravids, whose greatly elongated ribs must have supported a wing membrane permitting gliding flight from tree to tree.

Invasion of the seas

Archosaurs also invaded the sea. Fossil crocodilians such as the thalattosuchians of the Jurassic period had greatly elongated jaws resembling those of the modern gharial and were well suited to capturing fish. Some forms such as *Geosaurus* and *Metriorhynchus* had limbs transformed into paddles and a tail-bend which may have supported some fin-like structure. We do not know when the crocodiles' ancestors branched off from other archosaurs in the evolutionary tree, but they first appeared in the middle Triassic as terrestrial animals.

The shallow seas accommodated yet another group of reptiles, the Euryapsida, characterized by a single high opening in the cheek region of the skull. This group includes the placodonts, whose name refers to the large crushing tooth-plates on the palate and the hind parts of the jaws. Their front teeth were long and chisel-shaped, suited to pick up hard-shelled mollusks from the sea bed. Some placodonts developed body armor which superficially resembled the shell of a turtle. Also in this group were the small pachypleurosaurs, the larger nothosaurs (growing up to almost 4 meters, or 13 feet), and the plesiosaurs. Pachypleurosaurs and nothosaurs moved through the water by undulating, like an eel, whereas plesiosaurs had large limbs (paddles) which, supported by strong

▲ *With hollow bones, a reinforced spine, "warm-blooded" metabolism, and an insulative layer of hair, the pterosaurs are now believed to have been more efficient fliers than was first thought. Pteranodon soared far out to sea, feeding on fish caught in a pelican-like pouch. Its narrow wing membranes, supported by elongated fingers, spanned up to 7.5 meters (24½ feet).*

Ornithischia (with a bird-like pelvis). The Ornithischia were all plant-eaters. Some dinosaurs walked on two legs, although most used four, and they are well known for their often bizarre appearance. Some of the Saurischia eventually evolved into the largest meat-eaters that ever roamed the continents: the Carnosauria, such as *Allosaurus* and *Tyrannosaurus*. Others became the largest plant-eaters, such as *Diplodocus*, which reached a length of 30 meters (almost 100 feet), and *Brachiosaurus*, whose body, 20 meters (65 feet) long, stood 12 meters (40 feet) high.

Theropods, a group of saurischians, walked on two legs. They were agile predators, including not only one of the smallest dinosaurs known, *Compsognathus*, with an adult size comparable to that of a chicken, but also larger fast-moving animals. Some may have had quite complex brains, and nervous control of their physiology (including internal temperature regulation).

Taking to the air

Birds are thought to have descended from advanced theropods, and one of the links is *Archaeopteryx* which lived during the late Jurassic, 150 million years ago. Birds can therefore be viewed as the last surviving dinosaurs! Birds, however, are not the first vertebrates to have acquired the power of flight. The Pterosauria, first known from late Triassic fossils found in the girdles and a rigid trunk, functioned as hydrofoils much as in modern sea turtles.

Ichthyosaurs were the reptiles most perfectly adapted to living in water and resembled dolphins in their external appearance. The relatively small mixosaurs, rarely exceeding 1 meter (3¼ feet) in length, coexisted in great numbers with pachypleurosaurs in inshore areas, but some other species grew to a very large size. *Shonisaurus*, of the late Triassic, reached 15 meters (50 feet) and roamed the open sea.

THE HISTORY OF SQUAMATES

Lizards, amphisbaenians and snakes, which make up the great majority of modern reptiles, are squamates. Squamates are members of the subclass Lepidosauria. The origin of squamates must date back to at least the late Triassic, although the oldest fossils found are from the late Jurassic. By the end of the Cretaceous, there were squamates that we can classify as representatives of living families. However, a study restricted to France showed that modern species of reptiles, like those of amphibians, did not appear until the middle Miocene, about 20 million years ago. Because the sea is a barrier for most squamates, the geographical distribution of their various subgroups more or less reflects the pattern of plate tectonics—the slow drift of continental landmasses across the globe at the rate of a few centimeters a year.

By the Devonian period about 400 million years ago, the world's landmasses had drifted together to form a single supercontinent: Pangea. This was the world into which the first amphibians and reptiles evolved. In the Jurassic period, some 200 million years later, Pangea began to split into two large supercontinents: a northern one called Laurasia, which included most of modern-day Europe, Asia, and North America; and a southern one called Gondwana, which included modern-day South America, Antarctica, Australia, New Zealand, Africa, and India. Accordingly, most squamate subgroups can be understood as either gondwanan or laurasian in origin.

There was a short period in the late Cretaceous when intermittent land connections allowed animals to cross between North and South America, as they can today, but after the Cretaceous period North and South America remained separated until recent times. Iguanid lizards are an example of a group of southern origin which made its way from South into North America during a Cretaceous connection and, from there, dispersed overland to Europe, where the fossil iguanid *Geiseltaliellus* has been found.

The Atlantic Ocean separated North America from Europe during the early Eocene, but at the beginning of that epoch, 57 million years ago, the animals of the two continents still showed close affinities. (During that time there was a sea between Europe and Asia.) After the complete opening up of the North Atlantic, the European reptiles and amphibians evolved into types restricted to Europe, quite separate from those of North America, examples being the anguid lizards and an extinct group of varanid lizards known as necrosaurids.

Squamate fossils in Australia generally date back to about 30 million years, but the oldest of them are all representatives of typical gondwanan groups such as varanid lizards. Of particular interest is the giant varanid *Megalania* from the Pleistocene (2 million to 10,000 years ago), which reached a body size of 7 meters (23 feet) and weighed almost 600 kilograms (about half a ton).

The disappearance of *Megalania* and some other large species of reptiles coincides with the extinction of many large mammals toward the end of the Pleistocene, during the period from about 100,000 to 10,000 years ago. The causes of this "megafauna" extinction remain controversial. Some scientists favor the theory that climatic changes led to the mass extinction, whereas others believe that excessive hunting by humans had major impact on the number of species. The evidence remains inconclusive.

OLIVIER C. RIEPPEL

▼ *Ichthyosaurs, such as Ichthyosaurus, were the most aquatic of all reptiles, having a streamlined body and shark-like fins and tail. Well-preserved fossils have indicated that ichthyosaurs were live-bearers, and the young were born tail-first, as they are in modern dolphins and whales.*

HABITATS & ADAPTATIONS

Amphibians and reptiles occur throughout much of the world, even in the Himalayas and the Andes above 4,500 meters (15,000 feet). They are present on all continents except Antarctica and some reptiles can even be found on tiny, remote islands. They live in rainforests, woodlands, savannas, grasslands, deserts, and scrub; they occur on the ground, underground, in rock crevices and under debris, up trees, and in marshes, swamps, lakes, streams, ponds, and the sea. Some even take to the air briefly.

WHERE AMPHIBIANS LIVE

Only Antarctica, extreme northern Europe, Asia, and North America, and most oceanic islands do not have native amphibians. Caecilians are pantropical, but are absent from Madagascar, New Guinea, and Australia. Salamanders are widely distributed in temperate regions of the Northern Hemisphere, although one family, the Plethodontidae, is most diverse in tropical Central and northern South America. In contrast, frogs are found nearly worldwide, with the greatest diversity of species in the tropics—nearly the same number of species (about 90) occur at single sites in the upper Amazon Basin in Ecuador as occur in all of the United States.

Relatively few species of amphibians are aquatic (water-dwelling) as adults, although most of them have aquatic larval stages (for example, tadpoles in frogs). Most salamanders and caecilians, and about half of the species of frogs, are terrestrial (ground-dwelling). Some of these, especially the caecilians, spend most of their time below ground. A few salamanders and many frogs are arboreal (tree-dwelling).

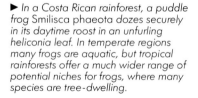

▶ In a Costa Rican rainforest, a puddle frog Smilisca phaeota dozes securely in its daytime roost in an unfurling heliconia leaf. In temperate regions many frogs are aquatic, but tropical rainforests offer a much wider range of potential niches for frogs, where many species are tree-dwelling.

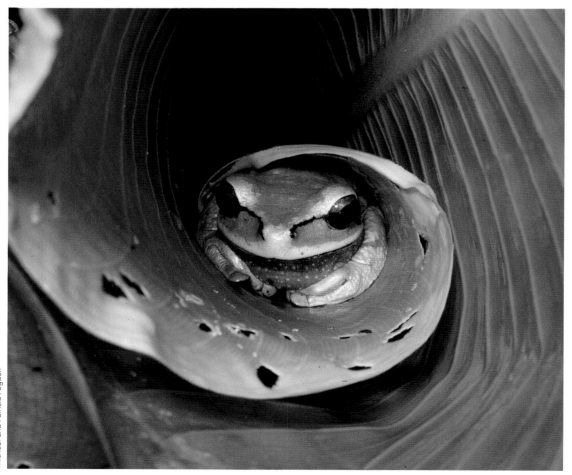

Michael and Patricia Fogden

Reg Morrison/AUSCAPE International

◀ Despite their dependence on water for at least part of their life cycles, a number of frogs have successfully adapted to deserts and other arid environments. One of the most bizarre is the water-holding frog Cyclorana platycephalas of the Australian interior. During drought it buries itself underground, shedding its loose, baggy skin and lining it with mucus; the skin dries and hardens to form a waterproof barrier, within which the frog lies dormant until the rains come again.

R.W. VanDevender

▲ Caecilians are amphibians that have evolved to suit an almost entirely subterranean life style. Adaptations include a heavily ossified (bony) skull, reduced eyes, and lack of any trace of a pelvic girdle. This is Oscaecilia ochrocephala of Central America.

AQUATIC AMPHIBIANS

Four families of salamanders in North America (Proteidae, Cryptobranchidae, Amphiumidae, and Sirenidae) are strictly aquatic, as are two cryptobranchids in China and Japan (the giant salamanders), and one proteid in southeastern Europe (the olm). These, and some North American plethodontid salamanders that live in subterranean waters, are neotenic; that is, they never complete metamorphosis from the larval stage; they include some of the largest salamanders, which breathe by means of gills and retain larval tail fins. The sirens (family Sirenidae) and congo eels (family Amphiumidae) are limbless, or nearly so, and swim in a serpentine manner. In contrast, the mudpuppies and olm (Proteidae) and the giant salamanders and hellbenders (Cryptobranchidae) have depressed bodies and normal limbs; they are more sedentary and crawl about on the bottoms of lakes or rivers.

Most of the so-called pond frogs of the genus *Rana* have powerful hind limbs and fully webbed feet; they are well adapted for leaping from land into water, where they are excellent swimmers. Practically all frogs are capable of swimming, but members of two families, in particular, are highly specialized swimmers. The paradox frog and its allies (family Pseudidae) in South America have extremely powerful hind limbs, huge fully webbed feet, and dorsally protruding eyes; these frogs are active near the surface of ponds. The clawed frogs of Africa and the Surinam toad and its relatives in South America (family Pipidae) also have fully webbed feet, but they have depressed bodies and relatively small eyes. They rest and feed on the bottoms of ponds and come to the surface

periodically to gulp air. If ponds dry up, these frogs burrow into the mud and remain there until the ponds fill with water again.

One subfamily of caecilians (Typhlonectinae) in South America is strictly aquatic. These eel-like amphibians live in lakes and rivers.

BURROWING AMPHIBIANS

While few salamanders are known to burrow in mud or soft soils, the caecilians are the most adept burrowing amphibians. These slender, worm-like, limbless animals have muscular bodies and compact skulls. Their eyes are covered by skin or bone, so they are blind; but with a protrusible, sensory tentacle lying in a groove or cavity on the head, caecilians are capable of detecting environmental cues and prey (mostly earthworms) by chemosensory means. Caecilians propel themselves through the soil by concertina locomotion, in which part of the body remains in static contact with the earth, and the adjacent parts of the body are pushed or pulled forward at the same time.

Frogs of diverse groups have a large, keratinized, spade-like tubercle on the inner edge of each hind foot. Spadefoot toads (family Pelobatidae) in North America, rain frogs (Microhylidae) in Africa, sand frogs (Ranidae) in Africa, the Central American burrowing frog (Rhinophrynidae), and the burrowing frog and Holy Cross toad (Myobatrachidae) in Australia, all use lateral scooping motions of the feet to bury themselves quickly in soft dirt or mud. A few frogs burrow head-first and have calloused pads on their snouts. African shovel-nosed frogs (Hemisotidae) have pointed snouts and burrow by

Michael and Patricia Fogden

▲ In general, amphibians require water for their early development. This presents some special challenges for those frogs that live as adults in trees but must somehow ensure the presence of water for their tadpoles. This challenge has been met in a variety of extraordinary ways; the female pygmy marsupial frog Flectonotus pygmaeus of South American highland rainforests, for example, carries her eggs in a moist pouch on her back.

These frogs walk about in trees by grasping branches. A few treefrogs (some *Agalychnis* and *Hyla* species in Central America and some *Rhacophorus* species in southeastern Asia) not only have huge, fully webbed hands and feet but also fringes of skin along the limbs. These modifications combine to provide an extensive surface area when the limbs are partially extended and the fingers and toes are spread apart; after leaping from high perches these frogs can glide considerable distances.

Some frogs have other adaptations for life in the trees. Some of the casque-headed treefrogs in the American tropics have the skin co-ossified with the underlying bones of the head and have a greatly reduced blood supply to this skin. During dry seasons, to reduce water loss, these frogs back into holes in trees or the "cup" of bromeliad plants containing small amounts of water; they flex their heads at right angles to the body and block the holes with their heads.

Many kinds of treefrogs mate and deposit their eggs on vegetation above water, into which the hatching tadpoles later drop. Others adhere their eggs to the walls of water-filled cavities in trees or deposit eggs in arboreal bromeliads. In this way, adults do not have to leave the trees to reproduce. The ultimate adaptation for arboreal reproduction occurs in some of the South American marsupial frogs (Hylidae), in which the fertilized eggs develop into miniature froglets in a pouch on the mother's back. Thus, it is possible for generations to pass without any individuals leaving the trees.

WILLIAM E. DUELLMAN

vertical motions of their heads. The Australian turtle frog and the sandhill frog (Myobatrachidae) have blunt snouts and dig with their forefeet.

Frogs may bury themselves to seek daytime retreats or more importantly for long periods of aestivation (in a torpid condition) during dry seasons, which may last for many months. During these times, frogs are subject to desiccation, so before aestivating they fill their urinary bladder with water. Some frogs, such as the African bullfrog (Ranidae), the water-holding frog (Hylidae) in Australia, and horned frogs and their allies (Leptodactylidae) in South America, reduce the risk of desiccation even further by making a cocoon. Once they are deep in their burrows the frogs shed the outer layer of their skin and secrete mucus, which together with the shed skin hardens to form a cocoon around the frog.

ARBOREAL AMPHIBIANS
The terminal segment of each finger and toe of many treefrogs (Hylidae, Centrolenidae, Hyperoliidae, Rhacophoridae, and a few Microhylidae) is expanded into a specialized toepad, by means of which these frogs are able to adhere to vertical surfaces. Many treefrogs also have slender bodies and long limbs, modifications that allow them to leap from one leaf or branch to another and to hold onto their perches. In one group of hylids (*Phyllomedusa* species) in the American tropics, the innermost fingers and toes are elongated and opposable to the outer digits.

WHERE REPTILES LIVE
Like amphibians, reptiles are ectotherms; that is, they do not produce very much of their own body heat and must rely on the external environment for it. For this reason they are sensitive to temperature and the number of species decreases toward higher latitudes and elevations, until eventually they drop out altogether. Nevertheless, one hardy species of lizard and one species of snake occur above the Arctic Circle in Scandinavia, and on some mountains lizards skirt around snow banks in their daily activities. The tuatara has been known to chase and catch seabirds on a night when air temperature was 7°C (45°F) and there was a driving rain and a wind of 50 knots! Such examples are unusual, however, and there are two kinds of places where reptiles really excel. They are abundantly tropical, and they form a conspicuous part of the desert fauna.

Turtles and crocodilians are mostly aquatic, whereas lizards and snakes are mainly terrestrial or arboreal. There are interesting exceptions, however: some tortoises not only live away from water, but they do so in desert regions, and some sea snakes live a totally aquatic existence.

TERRESTRIAL REPTILES

Because living on land is so familiar to humans, adaptations to other habitats often are considered "special" and terrestrial adaptations merely as the norm. However, terrestrial life imposes unique conditions that require modification of an animal's form and function just as much as does life in water or in trees. The skeleton needs to be strong enough to support body weight without the buoyant support offered by water. There needs to be a means of renewing the air over the respiratory surface (breathing) and a means of propelling the animal along the ground. Ability to see at a distance is important. Lungs, limbs, and eyes seem standard equipment to humans because we have them and we live on land, and because most of the animals we see have similar structures. In a total biological context, however, they are remarkable adaptations.

The limbs of reptiles are highly adapted to the kind of environment they occupy. Terrestrial lizards that run rapidly usually have relatively long legs with well-developed toes and claws that aid in gaining purchase on the ground. Some lizards with exceptionally long legs rear up and run on the hind legs only, thereby increasing their speed. These species have an exceptionally long tail, which serves as a counterbalance during such bipedal running.

Terrestrial snakes, entirely lacking limbs, have a different suite of adaptations. They are able to move across the ground by bending the body and pushing backward against irregularities of the surface. The large transverse scales on the underside are firmly attached in front, but have a free edge behind, where they overlap the scale following. These free edges catch against the ground and prevent backward slippage. (In sea snakes, which no longer need to have a grip on the ground, the belly scales are reduced in size, and in some cases are the same size and shape as the dorsal ones.) Many terrestrial snakes are long and narrow and have relatively long tails, attributes that are associated with rapid movement over the ground.

ARBOREAL REPTILES

Living in trees requires being able to grip trunks and branches to avoid falling, and being able to get from one branch to another. Many climbing lizards have sharp claws that can dig into bark and assist in climbing. Others, such as geckos, have expanded pads of lamellae on the toes, which allow them to grip almost smooth surfaces. Their effectiveness is obvious when you see a gecko run upside-down on the ceiling! Other arboreal lizards, such as chameleons, have toes that are opposable (like the thumb and fingers of a human) and can grasp twigs. Chameleons also have a prehensile tail; they can curl it around a branch and hold on, monkey-style.

The limblessness of snakes requires that they have different adaptations. Many arboreal snakes are amazing climbers, some even able to go up a vertical tree trunk without coiling around it; they merely utilize crevices in the bark for purchase. Some arboreal snakes, called vine snakes, look very much like vines. They are long and thin and drape over branches. This is an effective camouflage, but their length also assists in bridging gaps when moving among branches. Some arboreal snakes have a triangular shape in cross-section that gives greater strength and rigidity to a body extended, unsupported, across open areas.

Coping with a three-dimensional environment requires ability to judge distances. Many arboreal lizards and snakes have the eyes directed forward in such a way that both eyes can focus in front and achieve stereovision.

Perhaps the most unusual adaptation of arboreal reptiles is the ability to glide. A genus of Asian lizards (*Draco*) have extended ribs that support the normally loose skin of the flanks to form gliding "wings". If molested these "flying lizards" will escape out of their tree and glide long distances, either to the ground or to another tree trunk. There is also an Asian "flying" snake (*Chrysopelea*). It leaps from trees and by flattening its body can glide and break its fall. But there are no modern reptiles that match the soaring of the extinct pterodactyls.

▼ *Tree-dwelling snakes tend to be long and thin, like this plain tree-snake* Imantodes inornata *hunting among the twigs and foliage of a* Heisteria *tree in Costa Rica. A slender snake is better able than a fat one to span the gaps between branches, and many species are also triangular in cross-section, to combine to best effect the contrasting advantages of suppleness and rigidity.*

Michael Fogden

SUBTERRANEAN REPTILES

Some of the modifications of burrowing reptiles are not true adaptations, but rather loss of structures that no longer have a biological function. For example, in a dark burrow where the head is in direct contact with the soil, eyes serve no useful purpose, so some burrowing snakes and lizards have only rudimentary eyes. Limblessness also is usually a consequence of a burrowing existence. Although limbs can be useful in digging, and many terrestrial reptiles use them for digging burrows for shelter, they increase friction and require a larger burrow for animals that push through the soil, rather than dig. Accordingly, most truly subterranean reptiles have lost such encumbrances or have them greatly reduced in size.

By contrast, other features *are* adaptations to a burrowing life. Many burrowing lizards and snakes have the skull bones fused into a solid compact structure that can support the head as a battering ram. Blindsnakes (family Typhlopidae) have a sharp point on the tail, which serves as an anchor when pushing the smooth, highly polished body through the soil. Amphisbaenians employ another tactic: grooves around their body provide traction and they can "inch" forward in their burrows.

The limblessness of snakes is probably a heritage from their burrowing ancestors. It is believed that snakes arose from burrowing lizards that had reduced limbs and eyes. The snake eye lacks structures present in most vertebrates and appears to have been redeveloped from a rudimentary lizard's eye. Modern snakes have never regained functional limbs, and as indicated above, this condition has channeled their adaptation to

arboreal and terrestrial habitats along different pathways from those taken by lizards.

AQUATIC REPTILES

The terrestrial ancestry of modern reptiles has imposed limitations upon their life in water. Being egg-layers, females of most species must come to land to breed; only some of the marine snakes give birth to live young in the water and never voluntarily emerge onto land. Air-breathing is another limitation. All reptiles must come to the surface to breathe periodically, yet some have been able to prolong their diving time to a remarkable degree. Sea snakes can submerge for up to two hours, partly because they can absorb oxygen through their skins to a degree that land snakes cannot. They can dive to 100 meters (330 feet) and not suffer the bends, again perhaps because of skin permeability to gases; the excess nitrogen absorbed into the blood under pressure may pass through the skin into the sea and not build up enough to cause the bends. To keep air from leaking out of the lung, sea snakes have valves that close off the nostrils during dives. The mouth is tight-fitting as an extra precaution.

The nostrils and eyes of crocodilians and aquatic snakes tend to be located further onto the upper surface than in other reptiles. This allows an animal at the surface to ride low in the water, with fewer buoyancy problems, but at the same time be able to breathe and see.

Locomotion under water is completely different from moving on land, as pressure needs to be applied to a surrounding medium, not against a surface. Reptiles have adapted to this requirement in two ways. Those that have limbs have developed webbed feet or, in the case of sea turtles, flippers; and sea snakes have evolved a flattened paddle-shaped tail. Crocodilians and some semi-aquatic lizards also have the tail flattened, thereby increasing its surface area and propulsive thrust.

The sea poses problems in addition to those encountered by reptiles in fresh water. Reptilian kidneys cannot cope with high salinities, and life in the sea is possible by virtue of special salt-excreting glands. Sea snakes, sea kraits, and file snakes have one of their salivary glands modified as a salt-excreting gland. It is located beneath the tongue, and brine is excreted into the tongue sheath. When the snakes protrude their tongues the brine is pushed out into the sea. Another group of snakes inhabiting salty water, the homalopsines, have a similar gland, but located in the front of the roof of the mouth. Sea turtles have a modified tear gland that excretes brine from the eye, and the saltwater crocodile has small salt glands scattered over the surface of its tongue. Many terrestrial iguanas have a nasal gland that excretes excess dietary salt. The marine iguana, a large lizard from the Galapagos Islands, which

▼ *Reptiles as well as amphibians have many representatives that live almost entirely underground. Some burrowing snakes such as this Central American threadsnake* Leptotyphlops albifrons *feed mainly on ants and termites. Their sensitive sense of smell enables them to track foraging ants or termites back to their nests.*

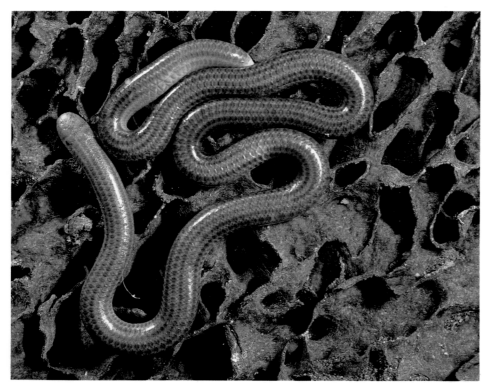

Gerald Thompson/Oxford Scientific Films

Stephen Dalton/NHPA

dives into the sea and feeds on marine algae, has this gland developed to a high degree. Brine is excreted into the nasal passage and then sneezed into the air when the lizard is on land.

Few animals are adapted to move over the surface of the water. One such rarity is a reptile, the basilisk, or Jesus lizard, from Central America. It has flaps of skin on the sides of the toes of the hind feet. These are folded up when it walks on land. If frightened it skitters bipedally out over the surface of a stream or pond; the flaps open up and provide enough extra surface area for it to run across the water, as long as it maintains momentum. If it stops running it sinks and must swim.

REPTILES ON ISLANDS
Remote oceanic islands usually have very few species of plants and animals, mainly those capable of surviving long voyages. Certain reptiles, notably geckos, are usually well represented, because several adaptations contribute to their success in dispersing to islands and their establishment once there.

Many lizards live in and under driftwood on beaches. Logs are washed out to sea and eventually drift to a distant shore, carrying with them lizards or their eggs ensconced in crevices. Some geckos have salt-tolerant eggs, which are sticky when laid but after drying adhere tightly to the sides of cracks and crevices.

A major problem colonizing animals face is establishing a population once a remote island is reached. Dispersal by rafting is a relatively rare phenomenon, and a second raft may not arrive at an island during the life-span of an individual reptile. Thus, most arriving species would die out with the death of the colonizing animal, unless such a miniature ark carried at least one individual of each sex. But parthenogenetic species (those in which the females can lay fertile eggs without being inseminated by a male) do not have such restrictions. It is probably significant that some of the geckos characteristic of remote islands are parthenogenetic, the males being either scarce or completely absent. In such species only one individual or egg need arrive at an island for a colony to become established. A widely distributed Pacific blindsnake is also parthenogenetic.

Perhaps the land reptiles with the record for overwater dispersal are the iguanas of Fiji and neighboring islands in the Pacific. Their only relatives are in South and Central America. Clearly, their ancestors rafted across the Pacific Ocean and, once isolated, evolved into the species of today.

HAROLD HEATWOLE

▲ *A combination of momentum, rapid leg movement, and fold-away flaps along each side of the toes enables the basilisk lizards of Central and South America, sometimes called Jesus lizards (genus Basiliscus), to run over the surface of water for some considerable distance.*

REPTILE &
AMPHIBIAN
BEHAVIOR

Reptiles and amphibians are the only land vertebrates unable to control their body temperature by internal, physiological means. By and large, this means that unless they regulate their body temperature behaviorally, it will simply rise and fall with the temperature of their immediate surroundings. This places enormous constraints on every aspect of their life. Hatching, growing, feeding, reproducing, escaping from predators, and even just being active depends on attaining and maintaining a body temperature that allows them to function normally. Consequently much of the behavior of reptiles and amphibians is aimed at temperature regulation, and this requirement underlies much of the complex and often bizarre behavior of individual species.

AMPHIBIAN BEHAVIOR

Because amphibians have permeable skin and, as a rule, are unable to regulate their body temperature by physiological means, many aspects of their behavior are dictated by the environment. Most amphibians are nocturnal and are active only when environmental conditions are sufficiently moist to prevent their bodies losing too much water by evaporation.

However, some amphibians exhibit water-conserving behavior and even behavior that regulates their body temperature. For example, diurnal (day-active) frogs living in cool climates often bask in the sun and thereby raise their body temperature before foraging for food. Some of them periodically enter water to acquire moisture and to lower their temperature. The North American bullfrog regulates its temperature by changing its position in relation to the rays of the sun. Other postural changes, evaporative cooling by mucous secretions, and periodic rewetting of the skin, all serve to reduce or stabilize the bullfrog's body temperature. Conversely, tiger salamanders, living in the Rocky Mountains of North America, and tadpoles in many parts of the world increase their body temperature by moving into shallow water on sunny days. However, almost nothing is known about the behavior of the secretive, burrowing caecilians.

FEEDING BEHAVIOR

Most amphibians simply sit and wait for prey to come within reach. A few heavy-bodied frogs, such as South American horned frogs, lure agile prey by waving the long toes on their hind feet. Other frogs, especially the small poison frogs in tropical America, actively forage for small insects

by day and find their prey by sight. A Central American burrowing frog (genus *Rhinophrynus*) apparently locates subterranean termite tunnels by their odor, then breaks open the tunnel, inserts its nose, and laps up the termites.

DEFENSIVE BEHAVIOR

Faced with potential predators, many amphibians feign death, present an enlarged image or the least-palatable part of the body to the predator, confuse the predator by changing the characteristic shape of the body, or change their color as a warning. In some cases they may even attack the predator. Salamanders that have concentrations of poison glands on their tail may lash it at predators; others have many poison glands on the back of the head and butt their head at the predator. Some Asiatic salamandrids are capable of protruding the tips of their ribs through the poison glands on their flanks, greatly increasing the chances of a predator encountering unpalatable secretions. Many salamanders exhibit caudal autotomy—that is, they can break off a portion of their tail—so if a predator grasps the tail, it will break off and continue to wriggle, thereby distracting the assailant while the salamander escapes.

Feigning death is widespread among frogs: some fold the limbs tightly against the body and lie motionless on their backs; others assume a rigid posture with the limbs outstretched. Toads commonly inflate their lungs, thereby puffing up the body and presenting a larger image to the predator, at the same time lifting their body off the ground and sometimes tilting it toward the predator. Some South American leptodactylid frogs (*Physalaemus* and *Pleurodema*) have large glands in the groin resembling brightly colored "eyespots", so that when a frog lowers its head and lifts its pelvic region, it looks like the head of a larger animal and presumably frightens off the would-be predator. When faced with danger, some African and South American treefrogs open their mouth (often displaying a brightly colored tongue) in an apparently threatening pose. Some large frogs actually leap and snap at predators. Many kinds of frogs emit a loud scream when disturbed or grasped by a predator; the noise may serve not only to frighten the predator but also warn other frogs of danger.

AMPHIBIAN REPRODUCTIVE BEHAVIOR

The most obvious reproductive behavior of amphibians is the sound made by frogs, especially the mating call made by males. Each species has a distinctive call, which is recognized by other individuals of the same species. Many frogs that live in seasonal environments have short breeding periods during which the males move to temporary ponds and vocalize; these calls usually are acoustically simple and serve only to attract

females. Frogs that are active for most of the year in tropical and subtropical regions commonly have more complex calls with both courtship and territorial components—a male advertises his territory acoustically to other males and also attracts females. Not all species of frogs call.

Salamanders are not known to use vocalizations in their sexual behavior, although some can produce barking or squeaking sounds. At least some terrestrial species mark their territories with scent trails, which are recognized by members of the same species. Each species of European newt (genus *Triturus*) has a unique, complex courtship behavior, in which the male must elicit a response from the female before he can initiate the next phase of courtship. Courtship takes place in water, and during the mating season males develop bright colors and elaborate tail fins. The male positions himself in front of the female and initiates a series of tail movements, stimulating her visually, by touch, and by smell. New World lungless salamanders engage in a tail-straddling walk in the water or on land. The male initiates

▼ *In a defensive threat display against a Central American toad-eater snake* Xenodon rabdocephalus, *a cane toad* Bufo marinus *inflates its lungs, lifts its body clear of the ground, and tilts toward its attacker. If further provoked, the toad can eject a spray of poisonous secretion from glands on its head.*

The only frogs that have elaborate courtship rituals are some of the aquatic tongueless frogs. In the South American genus *Pipa*, the male grasps the female around the waist while she performs a series of somersaults during which eggs are expelled; the male then fertilizes the eggs and sweeps them onto her back with his feet, where they implant and develop.

CARE OF EGGS AND TADPOLES

Most amphibians that deposit their eggs in water immediately abandon them. In species that deposit their eggs on land, however, it is usual for a parent to stay with the eggs, preventing marauders from stealing them. Many frogs take care of their young in more active ways. For example, females of the South American marsupial frogs transport eggs on their backs or in a dorsal pouch. Males of Darwin's frog in Chile pick up hatchling tadpoles from egg clutches on the ground and carry them in a vocal sac until they complete their development. Males of the hip-pocket frog in Australia collect tadpoles and transport them in pockets in their flanks. Females of the Australian gastric-brooding frog swallow fertilized eggs; the eggs hatch, and the tadpoles complete their development in the stomach, which has to turn off its digestive functions during brooding.

Observers have seen the parents of some South American pond frogs (genus *Leptodactylus*) protect their free-living tadpoles; the tadpoles move in schools, and the female remains with them. Females have even been seen leaping at potential predators of the tadpoles. Perhaps the ultimate type of parental care in amphibians is the feeding of free-living tadpoles by the mother. In some tropical American poison frogs, the mother transports individual tadpoles to bromeliad plants which hold water; she periodically returns to each bromeliad and deposits unfertilized eggs to provide nourishment for each tadpole. Similarly, females of some New World treefrogs deposit fertilized eggs in water cupped by a bromeliad or cavity in a tree and subsequently provide other batches of unfertilized eggs for food.

WILLIAM E. DUELLMAN

Michael and Patricia Fogden

▲ *The female strawberry poison frog* Dendrobates pumilio *of Central America transports her tadpoles, one by one, to tiny pools formed by rainwater in bromeliad plants. She will return later to lay unfertilized eggs in the water as food for her young.*

▶ *The female Surinam toad* Pipa pipa *carries her developing young about with her on her back, snuggled tightly into specially formed pockets in her skin.*

courtship by rubbing the female's head with his chin, secreting chemicals from glands on the chin which stimulate her. He then moves forward and the female straddles his tail, pressing her chin to the base of his tail. If the female does not follow closely, the male may swing around and violently slap her head with his chin. Courtship terminates successfully when the female follows the male until he deposits a gelatinous capsule with a cap of sperm (spermatophore), which she then picks up with the lips of her cloaca.

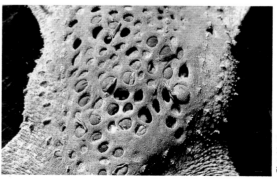

Chris Mattison

REPTILE BEHAVIOR

Reptiles have evolved into an impressive spectrum of shapes and sizes, which include shell-covered turtles and tortoises, long sinuous snakes, swift-moving lizards, and heavy-bodied crocodiles. They have therefore developed a wide variety of strategies to help them survive and reproduce. Most reptile species lay eggs, but some lizards and snakes give birth to living young.

Species recognition is critical. Individuals need to communicate with others of their own species so that they can get together to reproduce. This communication can be visual (as in displaying iguanid lizards), olfactory, including chemical substances called pheromones (as in most snakes), or vocal (as in lizards such as geckos, and in crocodiles and alligators). The signals provide not only information for species-identity but also the gender and reproductive readiness of an individual. Males in some species advertise their territory and perhaps their social dominance over other males using displays that are typical of their species.

TERRITORIALITY

In most families of lizards, males declare their territories by performing ritualized displays. Males in the Iguanidae and Agamidae families posture in ways that emphasize sexual differences in color and pattern, asserting to other males the ownership of a territory. These postural displays are reinforced by ritualized movements which, depending on the species, may be "push-ups" or bobbing of the head. Male marine iguanas may butt their heads together when one tries to claim the territory of another. In many monitor lizards, the males engage in combat by grasping each other with their front legs and, while standing high, try to push each other down. In some types of lizards a superior male asserts his dominance over other males with his displays, and the subordinate male assumes a submissive posture. Geckos, usually active at night, may declare this information with vocal signals. The territory provides an area for the adult male lizard to hunt for food and find a female.

Very little territoriality has been observed among snakes. However, in some families, such as the vipers and pit vipers, elaborate combat rituals take place between males—these may serve to establish dominance and the right to mate with a female. In the rattlesnakes' combat ritual, each male raises his head and body high, then twists against his opponent and attempts to topple him to the ground. In some tortoises, two males may butt their shells together in tests of strength to establish superiority.

COURTSHIP AND MATING OF REPTILES

During courtship the male performs certain behaviors, often ritualized and aided by pheromones, to attract the female and induce her

◀ A male banded anole Anolis insignis of Central America advertises his ownership of a territory by conspicuously displaying a colorful, sail-like dewlap on his throat.

Michael and Patricia Fogden

▲ *Plodding across a beach in Queensland, Australia, a female loggerhead turtle Caretta caretta returns to the sea at dawn after egg-laying at night. These and other marine turtles migrate vast distances across open ocean to reach traditional nesting sites, usually the same beach at which they themselves were hatched.*

to become sufficiently passive for copulation to take place. During the courtship of snakes the male usually crawls repeatedly over the female and orients his head to hers, at the same time bringing his tail into position next to hers to allow copulation to take place. In some species the male may hold the female by biting her head or neck region. In many lizards the male gets a biting grip on the female to hold her while he brings the base of his tail into position for mating. The male slider turtle courts his female by swimming backwards in front of her, stroking her face with his long fingernails. The male gopher tortoise circles his female on land, perhaps butting her shell, to induce her to become passive so that he can mount her from the rear. In North American box turtles, which have movable lower shells, the female may clamp her lower shell shut on the hind feet of the male, thus pinioning him as he falls on his back to mate; this position seems to be necessary because of the high-domed shells of these turtles.

REPTILES' PARENTAL CARE

In the typical nesting behavior of many reptiles, after mating the female digs a burrow, deposits her eggs, and leaves the eggs to hatch on their own. There is no protection for the young when they hatch, so many die and only a few survive to adulthood. In some species of snakes and lizards the female retains the eggs inside her body until the young are ready to be born; she gives birth to live young, which are on their own after emerging

from their fetal membranes.

Some female North American skinks stay with the eggs and brood them until hatching, even retrieving any eggs if a predator tries to make off with them. The females of some pythons coil around their eggs and incubate them; by contracting the muscles of her body, the mother can produce metabolic heat when needed to regulate the temperature of the eggs. Perhaps the most obvious nesting behavior is the mound building of the American alligator. The female then remains near the nest to protect it from intruders, and she may assist the young in hatching and escaping from the nest.

MIGRATION

Most reptiles do not migrate. For those that do, it is usually to reach a favorable nesting site or more favorable feeding sites. Prairie rattlesnakes may migrate up to 15 kilometers (9 miles) from their hibernation dens in the spring and then return to the same den in the fall. Marine iguanas of the Galapagos Islands migrate from the lava reefs to nesting beaches to lay their eggs, and then return to the reefs 100 meters (300 feet) away. To reach their nesting beaches, sea turtles may migrate 2,000 kilometers (1,240 miles). This involves sophisticated guidance mechanisms, which we do not yet understand. Certain populations of the green sea turtle migrate from the coast of Brazil to nesting beaches on Ascension Island in the mid-Atlantic Ocean, a distance of 5,000 kilometers (3,000 miles).

REPTILIAN DEFENSE

The most common type of defense in reptiles is biting. This is highly developed in many families of snakes, especially the true vipers, pit vipers, and elapid snakes (cobras and their relatives), which all produce venom. The only lizards that produce venom are the beaded lizards of the Americas.

The body structure of reptiles, and their habitat, affects the way they express aggressive behavior. Many lizards flatten their bodies and raise them high to appear larger, thus deterring a would-be predator. The bearded dragons and frilled lizard of Australia have elaborate throat fans, which they present to a predator, greatly enlarging their apparent size. Many snakes inflate or enlarge their neck in a threatening manner—as exemplified in some of the cobras. Many snakes and lizards threaten a predator by gaping their mouth, and some snakes flash a bright color on the underside of the tail. The North American hognose snakes respond to the threat of a predator by a sequence of unusual antics, first posturing as if to strike, then rolling over onto their back to simulate death, gaping the mouth, defecating, and becoming limp. If someone turns the snake over onto its belly, it immediately turns over again, belly up, as if to "play dead".

Cryptic coloration (camouflage) is common in reptiles, some of them blending with the habitat when they are motionless, while others have color patterns that produce a confusing appearance when in motion. Chameleons are freely able to change their color or color pattern as they move among different types of vegetation.

CAPTURING FOOD

The primary function of venom in a poisonous snake is to immobilize prey so that it can be devoured. Non-venomous snakes use other ways of obtaining food, such as coiling around prey to constrict it, as in the boas and rat snakes. Some reptiles actively pursue their prey, while others employ a strategy of "sit and wait" until prey comes close. The death adder of Australia may lie hidden in the leaf-litter with its tail exposed and a modified terminal scale intermittently waved from side to side as an insect-like lure to attract prey. The North American alligator snapping turtle lies on the bottom of a river with its mouth open, wiggling a small fleshy "worm-like" protuberance on its tongue to lure an unsuspecting fish.

CHARLES C. CARPENTER

▲ *Many snakes, like this African twig snake* Thelotornis capensis, *respond to threat by inflating the throat to exaggerate the apparent size of the head.*

◄ *A range of snakes and lizards "play dead" when threatened. An alarmed eastern hognose snake* Heterodon platyrhinos *of North America will roll over onto its back, defecate, open its mouth, and go limp. The response is so rigidly programed that, if rolled over onto its stomach, the snake will promptly "come to life" again, just long enough to roll over on its back and repeat the performance.*

ENDANGERED SPECIES

There have always been endangered species. Over periods of time measured in tens of millions of years, different lineages of species have evolved, expanded in numbers and importance, and then disappeared from the fossil record. By analyzing the different types of animals and plants fossilized in sedimentary layers from successive geological periods, we know that unrelated clusters of species have disappeared one after the other, during the course of millions of years, and probably for a variety of reasons. This happened to many amphibian groups during the Permian and Triassic periods (approximately 285 to 200 million years ago). But there have been times when a number of distinct groups completely disappeared in one relatively brief period, and perhaps with some common cause—as in the case of many reptile groups at the end of the Cretaceous, about 65 million years ago. It was during this period that all the dinosaurs then remaining became extinct.

► *A land iguana* Conolophus subcristatus *nibbles at a cactus plant. Confined to the Galapagos Islands, this species has become seriously endangered because of predation from dogs and cats released on the islands. It lives in burrows in sandy ground and feeds on plants and insects.*

K.H. Switak/NHPA

A NATURAL COURSE OF EVENTS?

Evolutionary theory predicts that species are likely to become extinct as one species out-competes another for some limited set of resources, such as the right type of food, or as a species becomes unable to adapt to a changing environment. In addition, over evolutionary time the genetic identity of species will change as populations evolve into what might eventually be a new and distinct species, or as some geographic barrier forces one species to split into two separate populations, each of which can embark on a new evolutionary career.

If both the evidence of the fossil record and evolutionary theory indicate that the decline and extinction of species is a natural course of events—that is, there have always been "endangered species"—why is there so much concern now about declining populations and contemporary extinctions?

The answer is that the rates of endangerment and extinction around the world are feared by many to be higher than ever before. In gathering up a disproportionate share of the Earth's resources and becoming the most successful single species the world has ever seen, humankind has modified virtually every component of the biosphere—from the deep ocean floor to the atmosphere itself. On a global scale, the rate and magnitude of these changes have increased sharply during the past few human generations, and continue to do so, as some people strive to meet the most basic needs of existence while others pursue a life style of excessive consumption. We have now observed the depletion and fragmentation of formerly widespread wild species, the diminution of species-rich habitats, and are able to compile lists of species seen 50 or 100 years ago but now apparently gone forever.

TWENTIETH-CENTURY EXTINCTIONS

No one should doubt that many species and their habitats face acute problems. We can predict that many sensitive species with small or now-fragmented populations will probably be lost. Yet it should be stressed that we actually know of few species that have been lost in our lifetime. The number of vertebrate species (those with a backbone) reliably known to have become extinct in the twentieth century is about 110. In purely numerical terms, this is of little significance compared with the 42,000 living vertebrate species known to science. The relative proportions are even more unbalanced among invertebrates and plants, but this has little meaning because the scientific inventory of these groups is far from complete, and evidence of extinction is very difficult to obtain.

Of course, it is impossible to know of all contemporary extinctions, and a significant number of species have certainly disappeared without ever having been collected, described, and named by science. This must be particularly true among invertebrates and plants, especially those inhabiting inaccessible areas.

FOUND JUST IN TIME

Some new species are named as a result of a taxonomist recognizing that a previously known single species is actually two, perhaps following new evidence of anatomical, biochemical, genetic, or behavioral differences between them. Other new species are really "new", in the sense of never having been recorded by science before. Sometimes these are discovered in circumstances that reveal how difficult it is for us to monitor the existence or status of such species.

For example, in 1981 the three-year-old son of a herpetologist was left at the roadside near a small area of forest in the Palni Hills of South India; he was to play at snake hunting while his elders got on with the real hunting nearby. The boy idly picked some cement out of a stone wall, and a small snake fell at his feet; the snake turned out to be a species new to science, eventually named *Oligodon nikhili*. Had that patch of forest been chopped down like the rest of the original forest cover, the species would almost certainly have disappeared even before its discovery.

Similarly, the anguid lizard *Diploglossus anelpistus* was described after only four specimens were found during the clearance of a forest area in the Dominican Republic. The species has never been seen for certain anywhere else on the island, and the forest area in which it was discovered has since been totally cleared. This species may now be extinct, although there are indications that it is nocturnal and semi-burrowing, so there is a possibility that additional populations remain

M.J. Tyler

▲ The gastric-brooding frog Rheobatrachus silus *inhabits mountain forest streams in eastern Australia. When its extraordinary breeding habits (the female incubates her young in her stomach and "gives birth" to them through her mouth) were first unraveled during the 1970s, it was reasonably common, yet during the 1980s it disappeared, for unknown reasons. Several subsequent thorough searches have failed to find it.*

▲ *Two butterflies imbibe eye and nasal secretions from basking yellow-spotted Amazon River turtles* Podocnemis unifilis *in Manu National Park, Peru. Inhabiting large rivers and wetlands in South America, several species of the genus* Podocnemis *have breeding habits somewhat similar to those of marine turtles, coming ashore to nest communally. Some have been brought to the brink of extinction by human exploitation of their eggs for food.*

facing a very high risk of extinction in the wild in the near future; these, together with species in less acute danger, are referred to as "Threatened Species". The terms are often used in somewhat different senses outside the IUCN system. Since the 1960s many regional, national, and provincial "Red Books" have appeared, produced by official bodies and by non-governmental organizations.

Despite its immense success, the Red Book approach does have limitations. Firstly, the biological assessment and monitoring that should precede inclusion of species on any "Threatened Species" list is most feasible in countries with high levels of scientific expertise, financial resources, and public concern. This inevitably leads to geographic imbalance in coverage: in the 1996 IUCN Red List of Threatened Animals, for example, the large number of species from North America and the small number from tropical countries is more likely to reflect the density of concerned observers than the number of threatened species. There is also considerable taxonomic imbalance, as some groups of animal are studied by a number of active specialists. Thus the conservation status of all species of turtles and tortoises, for example, has been assessed to some extent, and largely as a result of this interest there are many more turtles and tortoises in the IUCN Red List than there are snakes or salamanders. This is linked to the problem of size and visibility: far more is known about large species, such as crocodiles, turtles, and large iguanid lizards, than about small lizards, snakes, and salamanders, many of which are also highly secretive and seasonal in appearance. Furthermore, it is actually impossible to monitor the status of individual species in species-rich faunas, such as those of tropical rainforests.

hidden at other sites. The point is that there must be many cases where chance discoveries, such as the two mentioned here, simply do not happen, and therefore species are lost before their existence is known to science.

THE NEED FOR INFORMATION

There is still hope. Some of the trends mentioned can be slowed, some may be reversible, and our information base can be expanded. Armed with information and political strength, it may be possible for us to relieve and sometimes remove many of the pressures that reduce the world's biological diversity.

Information is a key requirement. The "Red Data Book" concept, originated during the 1960s by the late Sir Peter Scott, has been an important means of stimulating public involvement in conservation action. The intention was to catalog species thought to be, in varying degrees, in danger of extinction. Brief notes on distribution, threats, and conservation needs were included. A system of categories was devised to reflect the supposed imminence of extinction, and this has been most recently modified in 1996.

In the Red Data Book system of categories developed by IUCN (the International Union for Conservation of Nature and Natural Resources, now known as the World Conservation Union), "Endangered" has a relatively precise meaning and refers to species considered to be

"FLAGSHIP SPECIES"

Although attention is swinging away from individual threatened species, toward habitats and ecosystems generally, the great effort still made to protect individual species is not necessarily misdirected. Species such as the tiger and the giant panda are not only important in their own right, but they also serve as "flagship species"—that is, conservation and management of the extensive areas of habitat needed to support viable populations of these large animals will simultaneously improve the prospects of countless smaller animals and also plants.

There is no exact amphibian or reptile analog of the tiger, although protected areas set up in India for the gharial *Gavialis gangeticus* and the Indo-Pacific crocodile *Crocodylus porosus* (both threatened species) do include populations of other reptiles of conservation concern, notably freshwater turtles and the water monitor *Varanus salvator*. In general, conservation of amphibians and reptiles depends either on their incidental presence in protected areas which have been set up for other reasons or on individual-species action. The latter is sometimes the most appropriate course of action, because among the most-threatened species are crocodilians and marine turtles whose present predicament is the result of human exploitation—also an individual species-action.

STEPS TOWARD EXTINCTION

Although there is justifiable concern about the extinction of species, and perhaps a curious fascination about its finality, action can only be directed at species before they reach this stage. Some species may be beyond remedial action; these have been termed "proto-extinct". Others, while stressed to a greater or lesser extent, have the potential to respond to management interventions, and conservation action should therefore be directed at them rather than those probably beyond help. Information on the biology of the extinction process should assist in distinguishing one group from another.

A species can persist only when reproduction continues to be successful—when births are sufficient to offset deaths. This process can be disrupted in various ways. The direct exploitation of those individuals most reproductively active in a population will obviously reduce the birth rate. If carried on for a significant length of time— depending on the age when they first breed, and the age when they cease to breed, among other factors—the entire population may collapse. Large-scale disruption of habitats can reduce the availability of shelter and food, and therefore hinder reproduction just as effectively.

It has only recently become clear to what extent random events can affect populations, especially when populations have become numerically small and fragmented because of habitat loss or exploitation. Small populations are more likely than large ones to lose the genetic variability needed to meet new environmental challenges. Small and isolated populations are more likely to be eradicated by disease or natural disaster than large ones. In an extreme case, reproduction might cease because individuals of opposite sexes simply do not meet or have lost the social context in which migration or some other key breeding behavior usually occurs. Where fragmented populations are separated by large areas of now-unsuitable habitat, local populations that become critically depleted cannot be replenished by distant populations.

Species with these characteristics can in effect be doomed to extinction although individual members are still living. Among sea turtles, for example, virtually all the females that emerge to nest on a particular beach might be slaughtered, or all their eggs collected. There are beaches where this has occurred year after year. Because females are potentially long-lived, they still appear on the nest-beach every year, even though no young have entered the water for many years. If the period of intense exploitation exceeds the reproductive span of the females, the population will eventually disappear as the ageing females fail to be replaced by their progeny.

A prime example is one of the island populations of the Galapagos giant tortoise *Chelonoidis elephantopus*—the subspecies *C. e. abingdoni* in particular. This population now consists of a single male, found on Pinta Island in 1971 and held at the Charles Darwin Research Station since 1972. The tortoises on Pinta were heavily exploited by whalers and fishermen, and the vegetation severely degraded by the grazing of goats, introduced in 1958. Chance events doubtless played a major part in the reduction of this heavily stressed population to a single individual. While this concerns what is usually treated as a subspecies, rather than a full species, it is probably a good model of what has happened to extinct species.

▼ *Inhabitants of high mountain cloud forest in Central America, golden toads Bufo periglenes congregate in the breeding season to mate at small pools and streams. During the 1980s, this species (as well as many other frogs of highland forest around the world) suffered a catastrophic and largely mysterious decline, in many cases to apparently total extinction.*

Michael and Patricia Fogden

► *Feeding largely on lizards and frogs, the broad-headed snake* Hoplocephalus bungaroides *is confined to forests extending only a hundred miles or so around Sydney, Australia, where it is threatened by urban sprawl and degradation of its rocky habitat.*

J.C. Wombey/AUSCAPE International

THE EXTENT OF THE PROBLEM

It is impossible to assess just how many species are threatened at a global level. We can only deal with species that we actually know about, and we can only enumerate those that have been formally designated as threatened and listed in some organized format. The IUCN Red List is the most wide-ranging compilation of such species. Numbers of amphibians and reptiles included in the 1996 edition are given below, together with the number of each categorized as "Endangered". This table refers to full species only; it does not included the small number of geographic subspecies also in the IUCN Red List.

HABITAT LOSS

Habitat loss is the most common and pervasive cause of population decline and fragmentation; usually it is a complex of interacting factors rather than one simple one. Whatever immediate causes are involved, they are often the result of more distant social and economic factors operating at the national or global level.

Sometimes things are simple. The Israel painted frog *Discoglossus nigriventer* was described a few decades ago from specimens collected at Lake Huleh on the Israel–Syria border. Soon afterwards the lake was drained to allow agricultural development, and the species has never been

	Number of Species in IUCN Red List of Threatened Species	Threatened species that are "critically endangered"	Threatened species that are "endangered"		Number of Species in IUCN Red List of Threatened Species	Threatened species that are "critically endangered"	Threatened species that are "endangered"
REPTILES				AMPHIBIANS			
Turtles & tortoises	96	10	28	Frogs & toads	88	15	23
Crocodiles & alligators	10	4	3	Newts & salamanders	36	3	8
Tuatara	2	0	0				
Lizards	96	15	15				
Snakes	50	12	13				
TOTAL	254	41	59	TOTAL	124	18	31

seen since despite searches being made. It is presumed extinct.

When species are more widespread than the Israel painted frog appears to have been, disruption of habitat will more typically result in loss or depletion of local populations. This is the case with many amphibian species. In Europe, for example, countless small ponds and swamps have been drained to meet the demands of intensive agriculture, and countless local amphibian populations have disappeared along with them, simply as a result of the loss of their essential breeding habitat.

Less frequently, human activity provides new habitat for wild species and perhaps allows larger populations to persist than would otherwise have been the case. The rare and localized populations of the viperid snake *Vipera ursinii* in the mountains of southern Europe, for example, favor short grassland with abundant cover of juniper bushes—a habitat created partly by grazing sheep, under the traditional low-intensity system of subsistence farming. Now there tend to be fewer sheep, and the habitat is changing to the detriment of the snake populations.

More typically, human activity involves a piece-by-piece expansion into relatively unmodified habitats. Better roads lead to easier access, which leads to more people, more hunting or

disturbance, more construction, more forest clearance, more agricultural development, and so on. Narrowly distributed specialist-species are likely to be wiped out; wider ranging generalist-species will persist, but often in fragmented populations more susceptible to chance events, leading slowly to further population loss.

EXPLOITATION AND INTERFERENCE

Direct exploitation by humans is a cause of decline in relatively few animal species, but where it does occur its effects can be even more rapid and drastic than habitat modification. The traditional diet of many indigenous people often included several species of reptiles and their eggs, and sometimes the larger species of frogs. Although it was probably sporadic and moderate exploitation, it is difficult to establish how badly it affected wild species in prehistoric times. Today it is almost as difficult to distinguish between "subsistence" and "commercial" utilization. The latter, and international trade in particular, has caused severe decline in several species.

Frogs were collected commercially in the United States at the end of the nineteenth century. They are still collected in parts of South America and Europe, and the trade in frog legs collected in Southeast Asia and sent to Europe and elsewhere has now reached a high volume. Toad skin is

▼ *Amerindian youths with a captured anaconda (genus* Eunectes) *in Brazil. Anacondas are heavily persecuted by local human populations, and must be regarded as threatened. A few python species may grow longer, but anacondas are the largest and heaviest of snakes, sometimes exceeding 10 meters (about 33 feet) in length and 250 kilograms (550 pounds) in weight.*

Australian Picture Library/ZEFA

Michael and Patricia Fogden

▲ *The American crocodile* Crocodylus acutus, *here represented by a juvenile, remains fairly widespread over most of its former range, but its numbers have been severely reduced by hunting.*

▼ *A Pacific ridley* Lepidochelys olivacea *on a nursery beach in Santa Rosa National Park, Costa Rica. Scattered broken eggshells symbolize the fact that both this and the related Atlantic ridley are among the most critically endangered of the world's marine turtles.*

Mills Tandy/Oxford Scientific Films

processed into leather on a small scale.

Most species of crocodilians have been adversely affected by excess hunting for the skin trade, although the smallest species and those with too many bony plates in the skin have been less exploited. The worst period was during colonial times in the first half of the twentieth century, particularly as firearms became readily available during and after the Second World War. It became possible for hunters to cruise along waterways, detect the animals by torchlight (when the eyes give a distinctive glow in reflected light), and kill each one with a well-placed rifle shot. Species that frequented beds of dense aquatic vegetation were less affected, but entire breeding populations of less-secretive species were eradicated. The black caiman *Melanosuchus niger* of South America, the Nile crocodile *Crocodylus niloticus* in Africa, and the saltwater crocodile *C. porosus* in Asia, have all been major targets. Simple persecution or target practice also accounted for many crocodilians. Hunting, malicious persecution, and habitat loss because of drainage or dam construction, provide a powerful combination of threats. The precarious status of the Chinese alligator *Alligator sinensis* appears to be a result of such a combination.

Sea turtles have also been the victims of heavy trade pressures, with eggs, meat, shell, or skin variously taken as the main commodities. Nesting populations are especially vulnerable to excess hunting; it is theoretically possible for dedicated hunters to collect every female turtle that emerges during the nesting season. In practice, bad weather or sea conditions, or simple logistics, sometimes make this impossible—thereby

ensuring the continued survival of some populations that would otherwise become extinct. Nevertheless, many nesting populations of olive ridley *Lepidochelys olivacea*, the green turtle *Chelonia mydas*, and the hawksbill *Eretmochelys imbricata* have been irretrievably lost. Because females typically nest on the beach where they first nested and laid their clutch, it is almost impossible for lost nesting populations to be replaced by members of another population.

Larger lizards, including monitors (genus *Varanus*) in Africa and Asia, and tegus (genus *Tupinambis*) in South America, and also many snakes, such as boas, pythons, cobras, and vipers, are intensely exploited for the skin trade. Many amphibians and reptiles are in demand for the live animal trade. It is difficult to measure the effects of this activity on wild species, but there is little doubt that local populations can be adversely affected.

The accidental or intentional introduction of species into habitats where they were formerly absent has been a major problem, especially on small islands. The mongoose in the West Indies; goats and rabbits in the Mascarenes; cats, dogs, and pigs almost everywhere else; all have had detrimental impacts on indigenous species of animals. The initial effect might be caused by direct predation, as with the mongoose preying on lizards of the genus *Cyclura* and snakes of the genus *Alsophis* in the West Indies. Or the initial effect might be on habitats, as with goats and rabbits on Round Island north of Mauritius, where much of the original native plant-life has been destroyed, and at least one of the endemic reptiles has disappeared.

◄ Komodo dragons Varanus komodoensis *mating. Largest of all lizards, komodo dragons are confined to a few small islands in Indonesia. Although threatened, their notoriety as potential man-eaters and consequent high status as a tourist attraction offers some hope that their population levels might be sustained by careful management.*

Patrick Fagot/NHPA

CONSERVATION MEASURES

Exploitation tends to be a direct problem, although one with complex socio-political origins. For this it is theoretically possible to find direct solutions. One option is legislation prohibiting the exploitation, and the creation of national laws intended to meet the requirements of international conventions can also be effective. The Convention on International Trade in Endangered Species of Wild Fauna and Flora (CITES) is the most wide-ranging of these, and it has had significant success in reducing or preventing international commerce of threatened species listed in its Appendices. Crocodilians, sea turtles, and many other reptiles and amphibians are prominent in these lists.

For some species, notably certain crocodilians, trade in animals reared in captivity from eggs or young collected in the wild is allowed. This can let economic benefits flow to local people, and also provides a reason for protecting the habitat of wild populations needed to replenish farmed stock. This option is not feasible for species with lower-value products or those more difficult to raise in captivity.

Habitat conservation is a more complex problem. But in a growing number of countries, development projects such as roads and dams cannot legitimately be started before an investigation is made into the probable impact on wild species and habitats. Local authorities often have the power to protect small areas of habitat, such as farm ponds. In general, "Protected Area" designation is a powerful conservation tool, particularly when a nation has the resources to implement land-use policies within its protected

areas. This is the best starting point for conservation of large areas where habitats (intact or human-modified) and their species require management.

BRIAN GROOMBRIDGE

▼ *The crested iguana* Brachylophus vitiensis *of Fiji. Because of low population size and small recruitment potential, reptile species inhabiting oceanic islands generally have an even more precarious status than those species inhabiting continental land masses.*

Cliff and Dawn Frith/Bruce Coleman Limited

49

PART TWO

KINDS OF AMPHIBIANS

KEY FACTS

ORDER GYMNOPHIONA
• 5 families • 34 genera
• 156 species

SMALLEST & LARGEST

Grandisonia brevis of the
Seychelles archipelago
Length: up to 112 mm (4⅓ in)

Caecilia thomsoni of South
America
Length: up to 152 cm (5 ft)

CONSERVATION WATCH
■ Many species are known from
only one or a few specimens, but
this does not necessarily mean
they are rare or endangered.
Caecilians are difficult to find,
and few scientists have sought
them. For these reasons, very
little is known about their
distribution and abundance.

▼ *Most caecilians are so obscure and
little-known that they lack English
names. This is Dermophis mexicanus
of Central America.*

CAECILIANS

Caecilians are worm-like, mostly secretive, burrowing amphibians confined to the tropical and subtropical regions of the world. Although their name (pronounced "see-sil-e-an") means "blind", most caecilians have small eyes, which in some species are hidden beneath the bones of the skull. The order Gymnophiona is the least diverse and most poorly known of the three orders of amphibians.

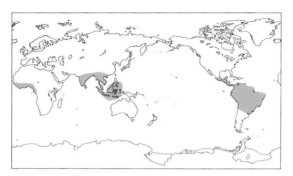

BUILT FOR BURROWING

Caecilians are the only living amphibians that are completely legless. They are capable of moving snake-like across surfaces when forced to do so, but normally they live underground and move through pre-formed tunnels or create new tunnels by pushing their head through loose mud or moist soil. Caecilians have many adaptations for burrowing. The skull is powerfully constructed with a pointed snout and an underslung lower jaw or recessed mouth, features that allow the head to be used as a ram. The eyes are reduced in size and importance, as there is no light in their underground world.

Caecilians have a unique pair of sensory organs called tentacles, one emerging from a groove or cavity on each side of the snout between the eye and the nostril. The tentacles, which are probably organs of taste and/or smell, are admirably suited for sensing the environment of tunnels. Another peculiar feature of caecilians is the manner in which they close their jaws; instead of having a single set of jaw-closing muscles as in all other terrestrial vertebrates (land-dwelling animals with backbones), caecilians have a dual mechanism that consists of two sets of jaw-closing muscles. Even this mechanism may be considered an adaptation for burrowing. Another interesting feature of caecilians is the presence of numerous skin folds, or rings, called annuli, which partially or completely encircle the body, although it is not

J. Campbell

Pavel German

yet known if the annuli are directly related to locomotion. Many species also have small fish-like scales hidden under the skin folds.

The body muscles of caecilians are arranged in such a manner that the body can act like a rod moving within a tube. The "tube" is the skin and outer layer of body muscles. The "rod" is the head, the vertebral column, and the deep body muscles associated with the vertebral column. With the tube fixed in position in a tunnel, the rod can be pushed slightly forward through the soil to extend the tunnel. When the tip of the rod (the head) has reached its maximum forward progression, the tube is pulled forward in waves and fixed in position again so that the rod can again be pushed through the soil. However, in very loose mud, caecilians use a different kind of locomotion; they simply swim eel-like through the watery medium.

REPRODUCTION

Unlike most other amphibians, all caecilians have internal fertilization; the males have a distinctive protrusible copulatory organ, called a phallodeum, which serves to inseminate the females. Like all amphibians, caecilians have eggs and embryos similar to those of fish, and unlike those of

reptiles, birds, and mammals. The latter three groups have a pair of special membranes (amnion and chorion) that enclose the embryos. Fish, caecilians, and other amphibians lack these membranes, and the embryos typically develop within a gelatinous egg case. Some caecilian species lay eggs; the eggs hatch into larvae; the larvae later metamorphose into adults. Other species are viviparous: they give birth to living young, and there is no larval stage.

Females of egg-laying species remain with their eggs until they hatch. The function of this parental care is unknown, but presumably the female protects her embryos against predation and perhaps other sources of embryo mortality such as destructive molds and desiccation.

Viviparous species have a number of unusual adaptations. The embryos developing in the female's oviducts soon use up all of the yolk provided in the egg, and the embryos then begin feeding on a substance called "uterine milk", which is secreted by the oviducts. Embryos are provided with numerous tiny teeth, which have distinctive shapes and are shed soon after birth. The function of these embryonic teeth is not understood, but they may have something to do with intrauterine feeding. The gills of embryos

▲ Most caecilians give birth to live young but some, like this sticky caecilian Ichthyophis glutinosus of southeast Asia, lay eggs in underground chambers. In these species the mother remains with her eggs until they hatch.

► Ichthyophis kohtaoensis *of southeast Asia. Most caecilians spend almost their entire lives underground, surfacing only in unusual circumstances. All appear to be carnivorous, feeding mainly on insect larvae and other invertebrates such as earthworms.*

R. Altig

developing in the oviducts are enormously developed in some species and are thought to function in gas exchange between the tissues of the female and her embryo. This has not been proved, and other functions are certainly possible.

FOOD AND ENEMIES
The food and feeding habits of caecilians have not been carefully studied, but the larval caecilians that have been dissected by scientists had eaten a variety of immature insects, earthworms, and other invertebrates. Land-dwelling caecilians seem to feed primarily on earthworms. Beetles and other insects have also been found in their digestive tracts, and occasionally small frogs and lizards.

Snakes, especially coral snakes, seem to be the primary predators of caecilians. In the Seychelles islands, chickens, pigs, and the shrew-like tenrecs (introduced from Madagascar) at least occasionally eat caecilians. None of these predators is native to the Seychelles.

Like all amphibians, the skin of caecilians contains numerous poison glands, which presumably function to discourage predators. Many frogs and salamanders with highly toxic skin secretions are brightly colored, and experimental evidence indicates that the bright

colors warn predators of the presence of the toxins and therefore help the amphibian to avoid injury from predator attack. Most caecilians have subdued colors, usually various shades of gray, but some (for example, members of the Rhinatrematidae and Ichthyophiidae families) have bright yellow lateral stripes. Although the function of the bright colors in these caecilians has not been studied, it seems likely that they too will prove to be warning colors.

PRIMITIVE CAECILIANS
Two families are considered to be relatively primitive, which means they are more like the common ancestor of all caecilians than are other caecilian families alive today. These are the Rhinatrematidae of northern South America and the Ichthyophiidae of India and Southeast Asia.

The rhinatrematids (two genera, nine species) of South America have retained the highest number of primitive characteristics: they have a "terminal" mouth (which means the mouth-opening is not recessed on the underside of the snout); their tentacles are in contact with the relatively large eyes; they have numerous skull bones; and they have a tail. These primitive caecilians also have the annuli (skin folds, or rings) subdivided into secondary and tertiary annuli which completely encircle the body, and numerous scales throughout the length of the body. The significance of secondary and tertiary annuli is still uncertain, and scientists know of their existence only by studying the embryonic and larval development of certain species. The specialized dual jaw-closing mechanism of caecilians is least developed in members of this family.

The life history and ecology of these seldom-seen caecilians is very poorly understood. However, we do know that they deposit eggs in soil cavities and that the eggs hatch into larvae with tiny external gills and gill slits. The larvae live in seepages and streams until they metamorphose into the adult form. The adults live in moist soil, leaf litter, and rotten logs.

The ichthyophiids (two genera, 37 species) of tropical Asia look very much like rhinatrematids, and until recently they were classified in the same family. They too have terminal mouths, numerous skull bones, a short tail, abundant scales, and primary annuli that are subdivided twice. Relatively advanced traits of the Ichthyophiidae include the position of the tentacle, which is forward of the eye, and a well-developed dual jaw-closing mechanism. The life history of ichthyophiids is similar to that of rhinatrematids.

Ronald A. Nussbaum

Although few clutches of eggs have been found, all of these were attended by a female.

TRANSITIONAL CAECILIANS

A small group of caecilians confined to India is placed in the family Uraeotyphlidae (one genus, four species). Uraeotyphlids are in some ways intermediate between the primitive and advanced

▲▼ *Primitive caecilians: a species of Epicrionops (above) and Ichthyophis kohtaoensis (below). Unique to caecilians, the tiny, inconspicuous sensory tentacles before the eye are just barely visible.*

Ronald A Nussbaum

caecilians. Primitive traits include the presence of many bones in the skull, a short tail, and numerous scales. Advanced features include a mouth-opening recessed below the snout, tentacles very far forward of the eyes, and the condition of the primary annuli—these skin rings are subdivided only once (instead of twice as in the primitive families), and some of the annuli at the forward end of the body do not completely encircle the body. The dual jaw-closing mechanism is well developed. Although very little is known about the life history of uraeotyphlids, the few observations suggest their way of life is similar to that of ichthyophiids and rhinatrematids, but with a shorter larval period.

ADVANCED CAECILIANS

The remaining two families of caecilians—the Scolecomorphidae and the Caeciliidae (including the subfamilies Caeciliinae and Typhlonectinae)—may be viewed as relatively advanced groups with specialized form, structure, and life history. Scolecomorphids and caeciliines live on land and are well adapted for burrowing in moist soil; Typhlonectines are aquatic or semi-aquatic. These two families share a number of advanced characteristics, such as a reduced number of skull bones, a recessed mouth, absence of tertiary annuli, and reduced number of scales, and they are tailless. Even so, the evolutionary relationships of these families to each other and to the more primitive families are still not fully understood.

Scolecomorphids (two genera, five species) occur in equatorial East and West Africa. These are among the most bizarre of caecilians. They have large tentacles placed far forward on the underside of the snout in front of the mouth. The vestigial eyes are attached to the base of the tentacles. In resting position, the eyes are under the skull bones, but when the tentacles are protruded the eyes are carried outside the skull along with the tentacles. Scolecomorphids are the only caecilians that lack stapes, the tiny bones that conduct sound vibrations from the environment to the inner ear. Other advanced features of this family include the loss of secondary annuli (only primary annuli that do not completely encircle the body are present); and, as is usually the case with caecilians that have lost their secondary and tertiary annuli, scolecomorphids do not have scales. The body ends in a small, blunt "terminal shield", which lacks annular grooves. The East African species (in the genus *Scolecomorphus*) are viviparous; that is, they give birth to living young. The eggs are fertilized internally and are retained in the oviducts of the female, where embryonic development takes place. The aquatic larval stage has been lost, and the young are born as fully terrestrial juveniles. Nothing is known about the life history of the West African species of the genus *Crotaphatrema*.

▶ *Siphonops annulatus of South America. Caecilians are confined to tropical and subtropical areas around the world. With their minute eyes and underslung mouths, it is often difficult at a casual glance to determine which end is which. Caecilians can be distinguished from all other amphibians by the ringlike folds or segments, called annuli, in their skins.*

Chris Mattison

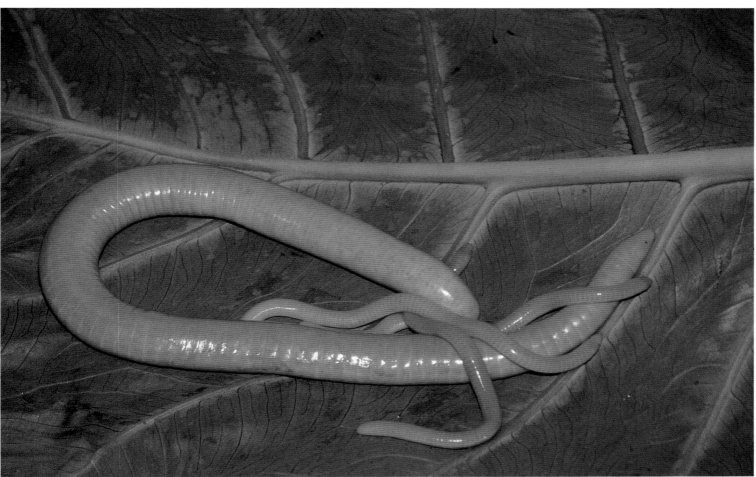

The Caeciliinae (23 genera, 89 species) is the most diverse and most geographically widespread of the caecilian subfamilies. Species of this subfamily are found in tropical Central and South America, equatorial Africa, the Seychelles archipelago in the Indian Ocean, and India. Caecilians are recognized largely by the condition of the skull, which consists of a characteristic number and arrangement of bones. The skull bones are relatively few, and they are strongly joined to transform the skull into a solid ram used in burrowing. The mouth of all caeciliians is recessed, with the exception of *Praslinia cooperi* of the Seychelles Islands, which has a terminal mouth. The position of the tentacle opening varies considerably among caeciliids; in some species, the opening is close to the eye; in others, it is closer to the nostril. Only in the enigmatic *P. cooperi* is the tentacle in the primitive position adjacent to and in contact with the eye.

Caeciliids have a variable number of their primary annuli subdivided into secondary annuli (these secondary annuli always occur at the tail end of the body); or they have only primary annuli. Similarly, caeciliids have a variable number of scales in the skin folds; and species that have no secondary annuli generally also have no scales. None of the caeciliids has a tail; in

some, the body ends in a blunt terminal shield.

Caeciliids have a variable life history. A few species, such as *Praslinia cooperi,* deposit eggs on land that hatch into water-dwelling larvae with a prolonged larval period. Others, such as *Grandisonia larvata* of the Seychelles Islands, have a very brief larval period. Still others, for example, *Afrocaecilia taitana* of East Africa, lay eggs on land and the eggs hatch directly into terrestrial juveniles without an aquatic larval stage. And some, like *Schistometopum thomense,* give birth to living young. Larval caeciliids are found in seepages and streams. The metamorphosed juveniles and adults are terrestrial burrowers. One species, *Hypogeophis rostratus* of the Seychelles, is found in streams, especially at night, as well as on land.

The subfamily Typhlonectinae (four genera, 12 species) is found throughout much of tropical and subtropical South America. Relatively primitive typhlonectines are semi-aquatic, whereas the more advanced species are aquatic. The skull of typhlonectines is basically like that of the burrowing caeciliids, with relatively few bones that are solidly conjoined and with a recessed mouth. Even the most aquatic typhlonectine species are adept at burrowing in soft mud and gravel in the bottoms and edges of aquatic

▲ *Viviparous caecilians, like* Schistometopum thomense, *give birth to small but fully formed versions of themselves. Many caecilians are brightly colored or boldly patterned, perhaps as a warning to predators of their poisonous skin secretions.*

Ronald A. Nussbaum

▲ Typhlonectes natans *of South America. Members of the subfamily Typhlonectinae differ from other caecilians in being aquatic or semi-aquatic rather than subterranean burrowers. Lacking tails or scales, they have laterally compressed bodies and a pronounced dorsal fin running the length of the back to facilitate movement through water.*

habitats. Typhlonectids have primary annuli only, which are sometimes wrinkled, giving the appearance of being divided into secondary annuli, and they have no scales. They are tailless, with a terminal shield that lacks rings. The aquatic species have a skin fold or fin that runs along the back from the end of the body toward the head; the body is compressed laterally, which together with the fin increases swimming ability. The tentacular opening of typhlonectids is very small and variously placed between the eye and nostril, but it is never in contact with the well-developed eye. Unlike other caecilians, typhlonectids apparently do not protrude their tentacles. One recently discovered typhlonectid is peculiar in that it has no lungs (all other caecilians have lungs), and the passages between the nostrils and the mouth cavity are sealed. Typhlonectids are viviparous, producing young that are fully metamorphosed at birth.

EVOLUTIONARY HISTORY

The origin of caecilians is unknown. Some scientists believe they are derived from microsaurs, a group of salamander-like amphibians that became extinct during the Permian period, about 250 million years ago. The fossil record of caecilians is very poor—in fact only two discoveries have been reported and it is difficult to draw any conclusions from these specimens. The first consists of a single vertebra from Paleocene deposits in Brazil; this 60-million-year-old fossil has been classified in the family Caeciliidae, suggesting that there has been very little change in this family over long periods of time.

The second discovery, which has not yet been fully reported, consists of several fossils from Lower Jurassic deposits in northeastern Arizona. These fossils, which are at least 170 million years old, have small but well-developed legs.

RONALD A. NUSSBAUM

IMPROVEMENTS FOR BURROWING

A typical vertebrate has a single set of jaw-closing muscles (one on each side of the head), which act by pulling up on the lower jaw from a point in front of the jaw hinge. Caecilians have this set of muscles too—labelled "adductor mandibulae" in the drawings. But they are unique in having a second set of jaw-closing muscles—labelled "interhyoideus posterior" in the drawings—which assist in jaw closure by pulling back and down on a special extension projecting behind the hinge of the lower jaw, in much the same way as pulling down on one end of a seesaw causes the other end to swing up. This set of muscles is present in other vertebrates also, but is not involved in jaw closure.

Why have these muscles been pressed in to service as jaw-closers in caecilians? The answer seems to lie in the evolution of burrowing efficiency. Relatively primitive caecilians, such as the rhinatrematid *Epicrionops petersi*, are relatively inefficient burrowers, which is reflected in both their behavior and their anatomy. Their mouth is at the terminal point of the snout, and they have a relatively weak skull, characters that would not be expected in a highly efficient burrower. In *E. petersi*, the adductor mandibulae muscles are large and dominate the jaw-closing system; they project up through openings in the skull, called temporal fossae, to attach on the top and side of the skull. The novel set of jaw-closing muscles—labelled "interhyoideus posterior"—while present and functional in these caecilians, is relatively small.

In somewhat more specialized burrowers such as *Ichthyophis glutinosus*, the temporal fossae are closed, making the skull more rigid. Closure of the temporal fossae restricts the size of the ancestral set of jaw-closing muscles by confining them under the roof of the skull. In *I. glutinosus*, the novel set of jaw-closing muscles has become dominant.

This evolutionary progression continues in forms such as *Microcaecilia rabei* and *Crotaphatrema lamottei*, in which the novel component of jaw closure is even more dominant. The highly specialized burrowing efficiency of these two species is indicated by the reduced number of skull bones, the rigidity of the skull, and the recessed or subterminal mouth.

FAMILY RHINATREMATIDAE
Epicrionops petersi

interhyoideus posterior

depressor mandibulae

eye

nostril

adductor mandibulae

opening for tentacle

FAMILY ICHTHYOPHIIDAE
Ichthyophis glutinosus

▲▼ *Caecilians are notable for the musculature of their jaws, incorporating a feature found in no other vertebrate. In most vertebrates, as well as some relatively primitive caecilians such as Epicrionops petersi and Ichthyophis glutinosus (above), jaw closure is primarily achieved by a set of muscles known as the adductor mandibulae. But in other caecilians, such as Microcaecilia rabei and Crotaphatrema lamottei (below, left), the jaw-closing function is dominated by the use of an entirely separate set of muscles, the interhyoideus posterior. The depressor mandibulae muscles are involved in jaw-opening.*

FAMILY CAECILIIDAE
Microcaecilia rabei

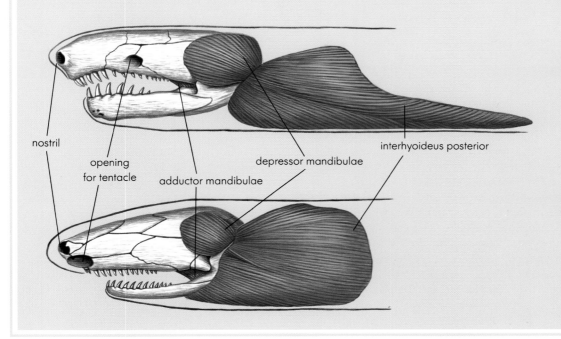

nostril

opening for tentacle

adductor mandibulae

depressor mandibulae

interhyoideus posterior

FAMILY SCOLECOMORPHIDAE
Crotaphatrema lamottei

SALAMANDERS & NEWTS

ORDER CAUDATA
• 10 families • *c.* 63 genera
• *c.* 440 species

SMALLEST & LARGEST

Lungless salamanders (genus *Thorius*) of southern Mexico
Head–body length: 14 mm (½ in)
Total length: 27 mm (1 in)

Chinese great salamander *Andrias davidianus*
Total length: 1.8 m (almost 6 ft)
Weight: 65 kg (143 lb)

CONSERVATION WATCH
!!! Lake Lerma salamander *Ambystoma lermaense*; desert slender salamander *Batrachoseps aridus*; and Sardinian brook salamander *Euproctus platycephalus* are critically endangered.
!! The 8 endangered species are: *Batrachuperus mustersi*; Abe's salamander *Hynobius abei*; Oki salamander *H. okiensis*; Hokuriku salamander *H. takedai*; *Batrachoseps campi*; Red Hills salamander *Phaeognathus hubrichti*; Jemez Mountains salamander *Plethodon neomexicanus*; Shenandoah salamander *Plethodon shenandoah*.
! 25 species are vulnerable.

S alamanders have been the subject of countless myths and legends since remote times. The origin of the salamander name is an Arab–Persian word meaning "lives in fire". In fact, until a few centuries ago it was believed that the black and yellow fire salamander *Salamandra salamandra* of Europe could pass unscathed through flames, a belief still held in some areas. Such modern names as the English "fire salamander" and the German "*Feuersalamander*" come from this ancient legend. In recent times salamanders have become a favorite of terrarium and aquarium keepers around the world because of the ease with which they can be raised in captivity.

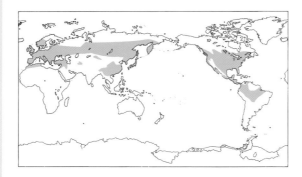

WHAT'S IN A NAME?
Salamanders belong to the order Caudata (from the Latin word *caudatus*, meaning "provided with a tail") and are also referred to as caudates; as the scientific name implies they retain a tail even after metamorphosis from the larval stage. In this they differ from other amphibians—caecilians have no tail or only a rudimentary one, and frogs and toads have a tail only in the larval stage.

"Salamander" is a broad term applicable to any member of the Caudata, whereas "newt" is more restrictive. Members of 10 genera *Cynops*, *Euproctus*, *Neurergus*, *Notophthalmus*, *Pachytriton*, *Paramesotriton*, *Pleurodeles*, *Taricha*, *Triturus*, and *Tylototriton*, which are mostly aquatic, at least during the breeding season, are commonly called newts; the others, whether amphibious or exclusively aquatic or exclusively terrestrial, retain the salamander name. "Newt" has a curious origin: it derives from the Anglo-Saxon word *efete* or *evete*, used to indicate newly metamorphosed newts, becoming *ewte* in Middle English, and then *newte*: an ewte = a newte.

Salamanders and newts are found almost exclusively in the Northern Hemisphere—in Europe, central and northern Asia, northwestern Africa, North America, and Mexico. In the Southern Hemisphere there are only a few members of the families Hynobiidae and Salamandridae in Southeast Asia, as well as the Chinese giant salamander, and species of lungless salamanders are found as far south as central Bolivia and southern Brazil. Caudates are found from sea level to about 4,500 meters (14,700 feet), wherever there is sufficient humidity during at least part of the year. Some specialized species dwell entirely in subterranean waters, others are tree and/or rock dwellers.

There are very few fossil remains testifying to the origin of this order. The earliest known true salamanders are three specimens dating back to the late Jurassic, about 150 million years ago. In 1726 the fossil remains of a giant salamander were curiously attributed by J.I. Scheuchzer to that of a human "witness to the Universal Flood", named *Homo diluvii testis*, and it was not until a century later that the remains were correctly identified and named *Andrias scheuchzeri*.

LIKE LIZARDS WITHOUT SCALES
Superficially salamanders resemble lizards but are immediately distinguished by their complete lack of scales. Even Linnaeus, the father of modern taxonomy, erroneously assigned some species of

▼ *Newts, like this crested newt* Triturus cristatus, *are common in freshwater streams and swamps of all kinds, although more or less restricted to the Northern Hemisphere. This male is eating its sloughed-off skin.*

Jane Burton/Bruce Coleman Limited

salamanders to the lizard genus *Lacerta*. Salamanders also resemble lizards in usually having four limbs; but again, like some lizards, certain salamander species have only forelimbs.

The body is lizard-like, with the head more or less distinct from the trunk, except in permanently aquatic species, whose head is elongated and rather eel-like; in this case the limbs are short, as in the olm *Proteus anguinus* of Europe and the American *Amphiuma* species, which some people call "congo eels", or the hind limbs are entirely absent, as in American sirenids.

The tail is laterally compressed and often crested in the aquatic species, and rounded or only slightly compressed in the terrestrial ones. In some species, particularly tree-dwellers, the tail is

also somewhat prehensile. When threatened by a predator, some lungless salamanders do what many lizards can do: voluntarily sever their tail (known as autotomy) as a decoy; the tail then regenerates. In most species the adults are 10 to 20 centimeters (4 to 8 inches) total length.

THE SKIN
Salamanders and newts continue to grow even after reaching sexual maturity, which is one of the reasons why the superficial horny layer of the skin is periodically shed. Depending on the species, the old skin comes off (as frequently as once a week in some cases) either in fragments or in one piece. This slough, or exuvia, is then usually devoured by the salamander.

▲ Perhaps from their natural propensity to crawl from crevices in logs tossed onto a campfire, salamanders have had a persistent association with fire in folk legends. It was widely believed that they cannot be harmed by fire. Occurring over much of central and southern Europe, as well as the Near East and northwestern Africa, this is almost certainly the species that originated in such myths, the fire salamander *Salamandra salamandra*.

► A salamander of the genus Bolitoglossa at night in a Costa Rican rainforest. Many newts and salamanders are nocturnal, but despite their obscurity and inconspicuousness, they are often very abundant. In some woods and forests, the total mass of these animals may outweigh that of all mammals and birds combined.

Peter Ward/Bruce Coleman Limited

Unlike that of reptiles, the skin of salamanders contains three different types of glands distributed rather uniformly over the entire body surface: mucous, granular, and mixed. The secretion of the mucous glands, and the mucous part of the mixed glands, is homogenous and frothy, neutral or alkaline, and sticky. On land its primary function is to protect the skin from drying out, permitting respiratory exchanges which could not occur through a dry surface. In water, it helps to maintain the body's internal osmotic pressure (the salt and water balance in the body fluids) and simultaneously acts as a lubricant during swimming. The granular glands, and the granular part of the mixed glands, produce a granular secretion containing various types of poisons and often giving off a specific odor; they are located especially on the upper part of the head behind the eyes (paratoid glands), on the tail, and on the sides of the back. The mixed glands, as the name implies, produce both a mucus and a granular secretion and are located over almost the entire body. Some members of the families Plethodontidae, Ambystomatidae, and Salamandridae possess a fourth type known as the hedonic glands, which resemble the granular and mixed glands and are located in various zones. Their secretion, containing pheromones, plays an important role during courtship and mating; particularly noticeable is the chin gland of most plethodontid males.

In some species the skin is smooth and in others it is bumpy, or even wart-like at the poison gland outlets. The skin of amphibious species is smooth when the animals live in the water but can become roughish when they live on land. The skin is rich not only in glands but also in pigment cells. In many species the body color is brownish, yellowish, or grayish, sometimes with a barely contrasting pattern. Several have gaudy coloration on their back or underside throughout the year in both sexes (as in the fire salamander and various lungless salamanders), or else only in the males, particularly during the season of courtship. The courtship period is also the time when many male newts develop conspicuous skin folds over their body, tail, fingers, and toes. Species that live permanently underground lack pigment and thus are white or pink. The color of the larvae usually differs from that of the adults.

With the exception of the two species of the Asian genus *Onychodactylus*, whose toes have a horny sheath which turns into a little blackish claw at least during the mating season, salamanders and newts do not have nails.

THE SKELETON
Some parts of the skull begin as cartilage and then turn into bone during growth. Others, not derived from cartilaginous elements, are known as membranous bones. The shape, number, and arrangement of these two types of bones are

important keys for classification.

The spine is generally divided into five regions: cervical, dorsal, sacral, sacro-caudal, and caudal. The cervical region consists of a single vertebra joined to the skull at four points (instead of two points in all other amphibians). The dorsal region usually has 13 to 20 vertebrae, although the minimum number is 11 (in the mole salamander *Ambystoma talpoideum*) and the maximum 63 (in *Siren* species). The sacral region consists of one vertebra; and the sacro-caudal from two to four. The caudal (tail) region varies from 20 vertebrae to more than 100 in some *Oedipina* species of Central and South America.

Those with an eel-shaped body, such as *Siren, Proteus,* and *Amphiuma,* have ribs corresponding with only the dorsal vertebrae closest to the head, but in other species the ribs occur along all or almost all the dorsal tract and sometimes also in the sacro-caudal region. In two types of newts, the sharp-ribbed newt *Pleurodeles waltl* and the alligator newt *Echinotriton andersoni,* the tips of the ribs can actually perforate the skin, increasing the likelihood that the newt's poison will enter the body of any predator.

The limbs have the basic structure common to all vertebrates; the front legs are linked to a thoracic girdle which does not articulate with the vertebral column, and the back legs to a pelvic girdle articulating with the single sacral vertebra.

Most species have four fingers on each forelimb and five toes on each hind limb.

THE DIGESTIVE SYSTEM

Salamanders have no salivary glands. Those that spend all their life in water have what is called a "primary" type of tongue, a fleshy fold on the floor of the mouth with very little mobility because it has no intrinsic muscles; it is not used to capture prey. All other salamanders have a fairly mobile and well-developed tongue, which is used—when they hunt on land—to procure food. In some lungless salamanders it is mushroom-shaped and can be darted onto the prey, just as chameleons do. The teeth, usually small, are implanted on the margins of both the upper and lower jaws, as well as on the roof of the mouth; the sirenids have them only on the latter. The teeth of some lungless salamander males have a sexual function (discussed later). Salamander larvae also have true teeth, in contrast to those of frogs and toads which have only horny structures.

The esophagus is fairly short and leads into the stomach with no regional differentiation and then an almost-straight intestine with little distinction between the small and large components. The terminal part of the rectum enlarges to form a cloaca, which contains the outlets of the urinary and reproductive tracts; its opening is located underneath the base of the tail.

▼ *The skin of amphibians as diverse as toads and salamanders contains glands that secrete toxic substances as a defense against predators. The sharp-ribbed newt* Pleurodeles waltl *of Iberia and northwestern Africa adds an unusual component to this defense: its ribs are long and sharp, and if grabbed by a predator, the tips of the ribs penetrate through the skin and its associated poison glands and, in effect, inject their toxins into the soft tissues inside the mouth of the attacker.*

R. Konig/Jacana AUSCAPE International

Stephen Dalton/NHPA

▲ *Often kept as a pet, the Mexican axolotl* Ambystoma mexicanum *is the best-known of the newts and salamanders exhibiting neoteny. It may retain gills and other larval features throughout its life, even as a breeding adult.*

RESPIRATION AND BLOOD CIRCULATION

The lungs of salamanders (in species that have them) are sac-like. Usually they are identical in size, although in some species the right lung is slightly shorter, and in amphiumids the left lung is rudimentary. The lungs of the spectacled salamander *Salamandrina terdigitata* and a few other genera, such as *Euproctus*, are tiny. All plethodontids and some hynobiids are lungless; respiratory exchanges occur only through the skin and through the mucous membranes of the mouth and throat. Respiration through the skin also plays an important role, at times essential, even in species with lungs, for instance when they hibernate underwater. The olm and other species that never lose their gills (known as perennibranchiates), as well as larvae, and neotenics (those species which occasionally retain larval features in the adult form—see page 505) can breathe either through their skin or through external gills which stick out in bright red tufts on the sides of the head. Only amphiumas have internal gills. The color of the gills is due to

the rich supply of blood, and respiration through the skin is enhanced by the marked vascularization of the skin.

The heart consists of a single ventricle in which the arterial and venous blood mix only partially. The left and right auricles are distinct in the lunged forms, and not completely separate in the perennibranchiate and lungless forms. The latter obviously do not have lung veins, and they have a smaller left auricle. Red blood cells are large (the eel-like amphiumas of the southeastern United States have the largest of all vertebrate animals), usually elliptical, and usually have a nucleus. According to the species, their number ranges from about 30,000 to 100,000 per cubic millimeter of blood; but as recently discovered in *Triturus* newts, the number of circulating red blood cells probably increases markedly also in other salamanders when environmental conditions prevent the blood from being sufficiently oxygenated.

The lymphatic system is fairly well developed. Salamanders have numerous lymph hearts under the scapula (shoulder-blades), at the root of the tail, and along the body flanks just under the skin. The lymph hearts are heart-like structures that improve the lymphatic circulation.

NERVOUS SYSTEM AND SENSE ORGANS

Salamanders have a relatively simple nervous system. The brain is rather small, but its front lobes are well developed, paralleling the marked development of the olfactory organs (detecting smell). In most species the eyes are large, with round pupils, but in permanently subterranean forms they are reduced in size and sometimes hidden under the skin. Species with well-developed eyes usually have upper and lower eyelids, plus a nictitating membrane known as the third eyelid.

There is no external ear, only a vestigial middle ear which does not have a cavity of its own. Vibrations are probably transmitted to the internal ear through the jaws or the bones connecting them to the skull, or else through the forelimbs and pectoral girdle. In the water, low-frequency vibrations are also perceived by lateral line organs (which also detect changes in water pressure and currents) found on the head and sides of the trunk in all larvae and in adults of many prevalently aquatic species. But only female spectacled salamanders, which enter the water to lay their eggs, develop lateral line organs that are active exclusively during this short period.

REPRODUCTION

Some salamander and newt species are exclusively aquatic, and some exclusively terrestrial. Others are clearly amphibious, and the females, or both sexes, return periodically to water

Jack Dermid/Bruce Coleman Limited

to reproduce. This usually occurs in the spring and can involve large migrations.

In the most primitive families, Crypto-branchidae and Hynobiidae, males fertilize the eggs outside the female's body. Fertilization is internal in other salamanders, although for sirenids observations still have to be confirmed. In species with internal fertilization the male deposits a gelatinous spermatophore (a capsule containing sperm) during mating, which can have a shape peculiar to the species or genus, with a

▲ *Amphiumas, like this one-toed amphiuma* Amphiuma pholeter, *are entirely aquatic, although they sometimes emerge on land during heavy rain. Lateral line organs, visible here as pitted areas on the head, perceive low-frequency vibrations as well as changes in water pressure.*

NEOTENY: A JUVENILE ALTERNATIVE

In many different groups of animals, including both amphibians and reptiles, some species achieve sexual maturity while still retaining many juvenile or larval characteristics. This phenomenon, known as neoteny, is common in amphibians, especially salamanders, where it may take a number of different forms.

Although not universally accepted, one classification of the various forms of neoteny is that of DuBois (1979):
Total neoteny: some individuals remain permanently in a larval state and never reach sexual maturity;
Temporary neoteny: delayed metamorphosis, often accompanied by large larval size, in some species of frogs and salamanders;
Partial neoteny or paedogenesis: sexual maturity and reproduction occur in animals

which are still in the larval stage of their life cycle. Among the amphibians, paedogenesis occurs only in salamanders and can be of three types:
Obligatory paedogenesis: metamorphosis cannot be induced even with hormone treatment, as in the European olm;
Quasi-obligatory paedogenesis: metamorphosis can be induced under certain environmental conditions, including hormone treatment. For example, the axolotl, a famous paedogenetic form of the Mexican mole salamander, will metamorphose if treated with the hormone thyroxin;
Facultative paedogenesis: the ability to achieve sexual maturity and reproduce for a long time before metamorphosis, as in some populations of the salamander genus *Triturus*.

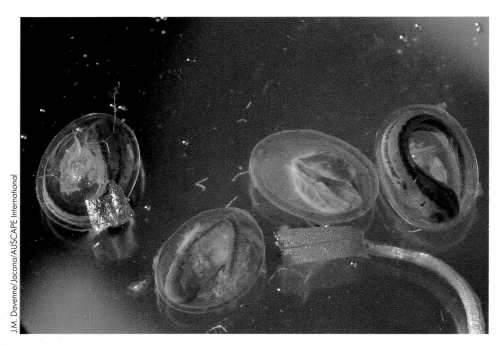

▲ *The alpine newt* Triturus alpestris *lays its eggs in water. Protected by a gelatinous translucent capsule, the embryos develop within, in a manner not unlike that of typical frogs.*

whitish mass of sperm adhered to its apex. The spermatophore, or only the sperm cap, is sucked in by the cloacal lips of the female.

Ovulation and fertilization may either follow insemination or be delayed for several months, in some cases up to two and half years, as in the fire salamander. When fertilization is delayed, sperm are stored in special diverticula of the female genital tract and subsequently fertilize the eggs at the time of ovulation.

Rutting takes place on land in exclusively terrestrial species and in those such as the fire salamander and spectacled salamander, whose

females go to the water only to lay their larvae and eggs, respectively. Rituals linked to courtship, such as nuptial dances, are fairly complex and differ from species to species. It can involve, as in *Euproctus* newts, a sort of embrace which is obviously not accompanied by copulation, as salamanders have neither a penis nor any other form of intromittant organ.

The eggs do not have a protective shell but are normally surrounded by a gelatinous layer whose development and consistency differ from species to species. On land eggs can be laid in several different places as long as there is sufficient humidity and protection. In water they are attached to rocks or submerged logs or roots. Some tree-dwelling lungless salamanders lay their eggs in bromeliads, in the water cupped in the base of the leaves. Some species, such as the olm, have parental care, a task usually carried out by the female but also by the male in the giant salamanders, the mud salamander *Pseudotriton montanus* and the clouded salamander *Hynobius nebulosus*. The number of eggs depends on the species and the size of the individual, varying from five to six in some terrestrial salamanders to a maximum of about 5,000 in some *Ambystoma* mole salamanders which lay their eggs in water.

Several salamanders are ovoviviparous— instead of laying eggs the female gives birth to well-developed larvae or to young that almost perfectly resemble the adults. The fire salamander usually gives birth to well-developed larvae but can in some regions give birth to already meta- morphosed young. In some ovoviviparous species adelphophagy takes place within the maternal genital tract; this consists of the more-developed

▶ *Crested newts in courtship. Caudates lack a penis or any similar organ, yet fertilization is usually internal. The male deposits his sperm in the form of a small packet, known as a spermatophore. Then, in an intricate courtship ritual, he maneuvers the female into a position to pick it up with the lips of her cloaca.*

embryos absorbing the eggs and smaller embryos, thus feeding on their siblings or "brothers" (*adelphós* is an ancient Greek word for brother). Many terrestrial lungless salamanders lay eggs that hatch already metamorphosed young.

The larvae usually have tail and dorsal membranes. In some species that reproduce in the water, before the limbs appear the larvae have a stick-like organ called the balancer or adhesive organ on the sides of the head in front of the gills, which serves to adhere the larvae to the substrate. After a period of time varying from a few days to several years according to the species, the more-or-less grown larvae undergo metamorphosis, losing their external gills and changing in other external and internal parts of the body. However, for environmental and/or genetic reasons, some or all members of a species can become sexually mature and capable of reproducing while maintaining such larval or juvenile morphological characters as external gills. The so-called perennibranchiate forms such as *Proteus* and *Necturus* never metamorphose, even if treated with a growth stimulant such as iodine or thyroid extract. Others do metamorphose spontaneously after a short time or when induced to do so when administered a hormonal treatment under experimental conditions. This is true for various species of *Triturus* newts and the axolotl *Ambystoma mexicanum*.

FOOD AND PREDATORS

Most salamanders are active at dusk and at night. When it is too cold or too dry many species take refuge under rotting vegetation or deeply buried rocks, in rock crevices or deep in the ground,

Hans Reinhard/Bruce Coleman Limited

▲ Several unusual features characterize the fire salamander's reproductive strategy. The female retains her fertilized eggs within her body until they hatch, and the emerging young may be either larvae (as in this case) or fully formed juveniles.

LIFECYCLE OF AN AMPHIBIOUS SALAMANDER

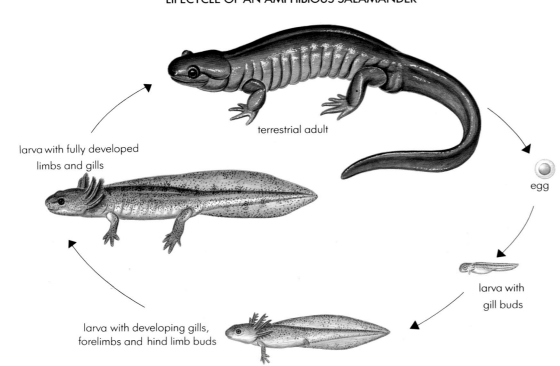

terrestrial adult

larva with fully developed limbs and gills

egg

larva with gill buds

larva with developing gills, forelimbs and hind limb buds

◄ Larval development in salamanders varies greatly between species. Illustrated is the basic amphibious lifecycle, in which much of the development occurs outside the egg. In many terrestrial salamanders, however, the young emerge from the eggs as miniature adults, with no aquatic larval stage.

becoming active only when the external environment is again suitable. They are all carnivorous and usually feed on small invertebrates such as insects, spiders, crustaceans, mollusks, and worms. The larger species also prey on small vertebrates. Cannibalism is not rare and may occur in both larvae and adults.

Unlike frogs and toads, salamanders do not usually make any sound as they have no larynx, or only a rudimental one, and no vocal cords. However, if disturbed or excited some species are capable of producing a weak squeak (the fire salamander, *Aneides*, *Dicamptodon*, and some species of newt), a faint yelp (sirenids) or a tiny shout (*Pleurodeles* newts).

Salamanders are preyed upon by other amphibians and by water tortoises, snakes, lizards, fish, birds, mammals, and large invertebrates. When in danger they defend themselves by secreting toxic or sticky substances, or by resorting to autotomy of the tail (as in some lungless salamanders) or by assuming a typical reflex posture which differs according to the species. One posture involves arching the body with the tail perpendicular to the body (or sometimes rolled up) to show off the gaudy ventral coloration that seems to warn the predator of a toxin in the skin. This particular posture is known as the unken reflex.

To humans the economic importance of salamanders is insignificant and is virtually limited to species of particular beauty and/or rarity, which are sought after as pets. Some species are used in biological research. The giant salamanders from Japan and China and the axolotl (the neotenic form of the mole salamanders) are appreciated as food, and the latter also as a supposed aphrodisiac. Others are used as live bait by fishermen, which has caused a great reduction in many populations of the lungless seal salamander *Desmognathus monticola* throughout the southeastern United States, and in some places have pushed them to the edge of extinction. The hynobid *Batrachuperus pinchonii* of eastern China is venerated as a divinity under the name "White Dragon", and the pool on Mount Omei where the species lives is the destination of pilgrimages, so much so that a sanctuary has been built nearby. This does not, however, prevent it from being dried and ground to a powder as a remedy for stomach disorders, a fate suffered also by its close relative, the Japanese clawed salamander *Onychodactylus japonicus*, which in traditional Japanese medicine has been used to rid the patient's body of worms.

Humans have had a direct impact on some populations through the capture of certain species for sale or study, and an even greater impact indirectly through the disruption or destruction of their habitat. Predation, especially on larvae, by fish stocked in pools for sport fishing has a very deleterious effect, but an even worse threat may come from acid rain.

THE MOST PRIMITIVE FAMILIES
The suborder Cryptobranchoidea includes the most primitive living salamanders, the only ones that have external fertilization. Their eggs are always deposited in two groups, each contained in a gelatinous sac. In the adults, two bones of the lower jaw (the angular and prearticular) are clearly separated from each other. In other respects the two families of this suborder, Cryptobranchidæ and Hynobiidae, differ greatly in appearance.

► *The hellbender* Cryptobranchus alleganiensis *is entirely aquatic, living in mountain streams in the eastern United States. Much the largest of North American salamanders (up to 70 centimeters, or about 28 inches, in length), it is nevertheless only about half the size of its relative the Chinese giant salamander, largest of all the caudates. Baggy folds of skin along the flanks increase the total surface area of skin, improving oxygen transfer from the water, but hellbenders also have lungs and periodically come to the surface to gulp air.*

R.J. Erwin/NHPA

Jack Dermid/Oxford Scientific Films

Hellbenders and giant salamanders

The cryptobranchids (family Cryptobranchidae) always live in running water. They are corpulent and have large skin folds along the flanks, which increase the body surface and thus enhance the absorption of oxygen from the water. Their metamorphosis is incomplete, so the adults still have gill slits or grooves and no eyelids. All species have four fingers on each forelimb and five toes on each hind limb.

The hellbender *Cryptobranchus alleganiensis,* which can reach a total length of 75 centimeters (30 inches), inhabits central to northeastern United States. The genus *Andrias* is today represented by the Chinese giant salamander *A. davidianus,* growing to 180 centimeters (almost 6 feet), and the Japanese giant salamander *A. japonicus,* whose length never exceeds 150 centimeters (almost 5 feet). They feed on various invertebrates such as crustaceans but occasionally also eat small aquatic vertebrates. Each female lays up to 450 eggs, in paired rosary-like strings. The male releases his sperm on the eggs and seems to guard them until they hatch 10 to 12 weeks later.

Hynobiids

In the family Hynobiidae none of the 36 species grows larger than 25 centimeters (10 inches) total length. They live in Asia, with the exception of the Siberian salamander *Salamandrella keyserlingi,* which has spread westward as far as European Russia. The family consists of seven genera: *Batrachuperus, Hynobius, Liua, Onychodactylus, Pachyhynobius, Ranodon,* and *Salamandrella.* They all reproduce in water (for example, streams, ponds, and tarns) but outside the breeding season are terrestrial. The female lays two spindle-like capsules, each containing 35 to 70 eggs, according to the species. Observers have often seen examples of parental care.

Some *Hynobius* and *Salamandrella* and all the *Batrachuperus* species have four-toed feet. Other types have five toes on each hind limb.

THE EIGHT ADVANCED FAMILIES

The suborder Salamandroidea, which includes the most advanced salamanders, differs from the primitive families in having the angular bone fused with the prearticular bone in the lower jaw, at least in the living species. With the probable exception of sirens, the sperm are taken into the female's body to fertilize the eggs.

There are eight families in this suborder, with some 50 genera and a total of about 380 species. Members of this suborder occur everywhere salamanders exist: all regions of the world except Antarctica, Australasia, Oceania, and Africa south of the Sahara.

Sirens

The family Sirenidae has so many peculiar characters that some authors classify it in a separate suborder, Sirenoidea, or even a different order, Trachystomata. Being neotenic (see page 505), sirens have gills throughout their lifetime, and they lack eyelids; they have small eyes, no hind limbs and an eel-shaped body.

Sirens live in southeastern United States and adjacent regions of Mexico, spending their life in ponds and swamps with rich aquatic vegetation and a muddy bottom; like eels, they can cover short distances on land at night during rainy periods. The greater siren *Siren lacertina,* which

▲ *Nearly one meter (3 ¼ feet) long, the greater siren* Siren lacertina *has a slender body resembling that of an eel; it has forelimbs but no hind limbs, and is equipped with lungs as well as gills. Sirens are exclusively North American.*

▲ *Two fully aquatic salamanders: the mudpuppy* Necturus maculosus *(top) and the dwarf siren* Pseudobranchus striatus *(above).*

▼ *The axolotl is normally muddy gray in color, but albinism, as in this pinkish individual, is not unusual.*

grows to 95 centimeters (37 inches) total length, and the lesser siren *S. intermedia,* reaching 68 centimeters (27 inches) have three pairs of gill slits and four fingers on each hand. The two dwarf sirens, genus *Pseudobranchus,* which grow to 25 centimeters (10 inches), have only one pair of gill slits and three-fingered hands.

Little is known about their reproductive biology, but judging by their uro-genital apparatus, one may suppose that they have external fertilization; their eggs have been found isolated or in small clumps attached to submerged plants. Like the lungfishes, if their habitat dries they can survive for weeks or months embedded in the mud, enveloped by a kind of mucus cocoon with only the tip of the snout jutting out.

Waterdogs, the mudpuppy, and the olm
Proteus, the sea god who had the power of assuming whatever shape he pleased, inspired the Viennese naturalist Laurenti in 1768 to use the name *Proteus anguinus* for a curious amphibian discovered 24 years earlier near Ljubljana (now in Yugoslavia). Its discoverer, Baron J.W. Valvasor, had considered it to be the juvenile form of an animal destined to change into a dragon. The nobleman was actually not far from the truth, as the little animal, fortuitously swept above ground by the flood of a subterranean river, really looked like a larva with its large red gill tufts. This was not only the first true dweller of caves and underground passages to be discovered, but also the first member of the small but extremely interesting salamander family, Proteidae.

Proteus, also known as the olm, can grow to a total length of 33 centimeters (13 inches) and is found only in the subterranean waters of the western Balkan Peninsula and northeastern Italy. It is eel-like and usually a pale rose color because there is almost no skin pigmentation. The eyes are small and concealed under the skin. The forelimbs and hind limbs bear three and two digits, respectively. Larvae and juveniles, darker and with larger exposed eyes, are noteworthy for feeding on bacteria, protozoans, and organic matter contained in the slime.

The family includes five other species, all in genus *Necturus*, which inhabit central and eastern United States. Because of the popular misconception that they can bark, four are called waterdogs, and these do not exceed a total length of 28 centimeters (11 inches); the fifth is the mudpuppy *N. maculosus*, sometimes a little more than 40 centimeters (15¾ inches) long.

All members of the Proteidae family are neotenic and consequently never lose their gills; they have two gill slits. They also lack eyelids and upper jaws. The eggs, which are tended by the parents, are attached to submerged stones or logs.

Three families of mole salamanders
The mole salamanders spend their life almost entirely underground, emerging from their subterranean world only to reach the ponds or streams in which they reproduce. The breeding season is usually spring, but at least one species, the marbled salamander *Ambystoma opacum*, reproduces during fall. Fertilization, often

preceded by a nuptial dance, is immediately followed by the laying of up to 200 eggs, generally in water but alternatively buried where rising water will flood the nest.

The family Ambystomatidae comprises about 30 species in the genus *Ambystoma*; they are found from southern Alaska and Canada throughout the United States and most of Mexico. Pacific mole salamanders are members of the family Dicamptodontidae, with three species in the genus *Dicamptodon* and of the family Rhyacotritonidae, with four species in the genus *Rhyacotriton*, all confined to northwestern United States.

These three families differ in that two (Dicamptodontidae and Rhyacotritonidae) have a lacrymal bone, while Ambystomatidae members do not. They also differ in their habitat preferences during the breeding season: the former prefer mountain brooks, while the Ambystomatidae prefer still water such as lakes. Some species, such as the axolotl, are neotenic (see page 65).

Two species, *Ambystoma laterale* and *Ambystoma jeffersonianum*, hybridize in eastern North America. It was thought that this hybridization had produced two all-female species that reproduced by parthenogenesis. However, later

▼ The Pacific giant salamander Dicamptodon ensatus *(top) is restricted to western North America. Unlike most salamanders, which are silent, this salamander frequently emits low-pitched sounds when disturbed. With a maximum length of 40 centimeters (15 ¾ inches), the tiger salamander* Ambystoma tigrinum mavortium *(bottom) is one of the largest terrestrial salamanders, and adults often feed on small vertebrates. It is highly variable in color and pattern, with spotted varieties as well as banded forms.*

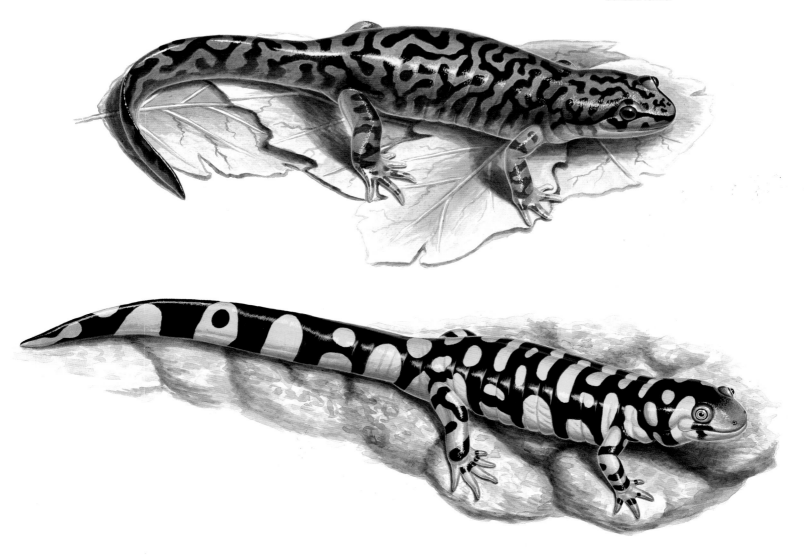

► Like this European crested newt (opposite), the skin of many species of newts and salamanders carry bold patterns in striking colors, signalling to potential predators the warning of poison glands in the skin. These patterns are sometimes on the throat or belly, and are often exhibited in a gesture unique to caudates, the unken reflex. In this rigid posture, the head is thrown back, the spine arched and the tail held stiffly upright, the better to display the visual threat.

evidence indicates that hybrid populations are maintained by breeding with one of the diploid bisexual species, and the hybrid offspring are the result of gynogenetic (that is, they devlop without the male's chromosomes entering the genetic complement of the new individuals) or hybridogenetic reproduction. The resulting hybrid populations are thus quite genetically diverse and not self-maintaining, so do not meet the criteria for distinct species.

All mole salamanders have a rather squat body, less than 35 centimeters (13 inches) in total length. Some have bright color patterns, which contrast with the dark color of the ground.

Amphiumas

If it were not for the presence of four tiny limbs, the amphiumas of the southeastern United States could easily be mistaken for eels, especially as they are neotenic and thus live mostly in water and, like eels, burrow in the mud; they can also move across wet ground. Adults have one pair of gill slits and inner gills, characters that increase their resemblance to eels. They have no eyelids or tongue. They lay their eggs under different kinds of shelters or on wet mud, in long strings each containing up to 150 eggs or more. The female remains coiled around them until they hatch, which takes about five months. On hatching the larvae must make their way to bodies of water, usually when it rains.

The family (Amphiumidae) includes three species: the one-toed amphiuma *Amphiuma pholeter*, at 30 centimeters (12 inches) total length; the two-toed amphiuma *A. means*, at 116 centimeters (46 inches); and the three-toed amphiuma *A. tridactylum*, at 106 centimeters (42 inches).

Salamandrids

The family Salamandridae includes some 60 species in 15 genera. Members of 11 genera more or less linked to water are commonly called "newts" (the scientific names are listed at the beginning of the chapter). The other four genera, predominantly or exclusively terrestrial, have the popular name "salamanders" in English, and are found in Europe, northwestern Africa, and southwestern Asia. The gold-striped salamander

Chioglossa lusitanica is endemic to Spain and Portugal, while the spectacled salamander is endemic to peninsular Italy. The genus *Mertensiella*, with two species in Turkey, some of the Greek islands, and the Caucasus, has males that bear a spur on the tail base. In the genus *Salamandra* there are two mountain-dwelling species, which give birth to perfectly metamorphosed young—the black, or black and yellow Alpine salamander *S. atra* from northern Albania to the western Alps, and Lanza's salamander *S. lanzai*, always black, endemic of the southwestern Alps—as well as the largest member of the family, the fire salamander *S. salamandra*. This magnificent black and yellow animal, sometimes longer than 30 centimeters (12 inches), has different subspecies throughout most of Europe extending to Iran and northwestern Africa.

Among the newts only three species of *Notophthalmus* and three of *Taricha* inhabit North America; the black spotted newt *Notophthalmus meridionalis*, the striped newt *N. perstriatus*, and the eastern newt *N. viridescens*, inhabit eastern North Amercia, while the larger (up to 22 centimeter or 8¾ inches) rough-skinned newt *Taricha granulosa*, red-bellied newt *T. rivularis* and Californian newt *T. torosa* may be found only in the marginal western United States. Two *Pleurodeles* species occur in northwestern Africa, with *P. waltl* also in Spain. All the other newts, about 45 species, live in Europe or Asia. The distribution of the genus *Euproctus*, sometimes referred to as brook salamanders, is interesting to biogeographers: while *E. asper* occurs in the Pyrenees mountains between France and Spain,

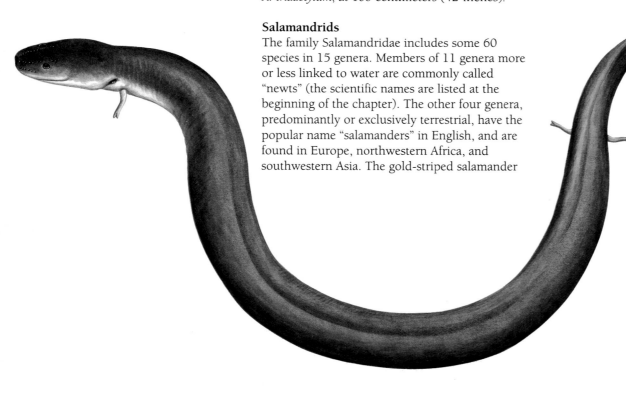

◄ The three species in the family Amphiumidae are identified by, among other things, the number of toes on their tiny limbs, and are appropriately named one-, two-, and three-toed amphiumas. They are restricted to the southeastern United States, where they are widely known as "Congo eels" — a confusing name because they are amphibians, not fish. This one is the two-toed species Amphiuma means.

▲▶ *Newts in the terrestrial phase, like this marbled newt* Triturus marmoratus *(above), take on the appearance of typical salamanders. The aquatic phase, as represented by this male alpine newt* Triturus alpestris *(right), is characterized by fin-like extensions on the tail and back, and sometimes the toes.*

E. montanus is endemic to Corsica and *E. platycephalus* to Sardinia, both islands of continental origin which detached from the south of France about 20 million years ago.

Only a few species are dull in color. Usually these salamanders and newts have lively and contrasting colors, some of them in gaudy patterns. Especially famous for the bright pattern of the breeding males are the 12 *Triturus* newts, and above all the marbled newt *T. marmoratus*, the alpine newt *T. alpestris*, and the extraordinary banded newt *T. vittatus*. The male banded newt develops a vertically striped and incredibly high dorsal crest. With the exception of a few species that give birth to live young, the Salamandridae lay their eggs in water. In *Triturus* and some *Cynops* newts the female secures each egg in the cavity of a leaf, which she folds between her hind limbs, ensuring that the egg is protected and oxygenated. Neoteny occurs also in the newts.

Salamandridae differ from the other salamander families in several skeletal characters too complicated to be discussed here, but they always have four well-developed limbs, with four fingers and generally five toes; only the spectacled salamander and some individuals of the newt *Echinotriton andersoni* have four-toed feet. They range in size from 7 to 30 centimeters (2¾ to 12 inches) in total length.

Lungless salamanders
Plethodontidae, the largest family of the order Caudata, includes more than 60 percent of the known living species—about 270 in some 30 genera—which are found in the Americas from Nova Scotia and southeastern British Columbia to central Bolivia and eastern Brazil. Only a few species are European, inhabiting Sardinia, the southeast of France, and Italy from the Maritime Alps to the central Apennines.

After metamorphosis they may be distinguished from all the other salamanders and newts by having a special structure, the nasolabial groove, which extends from the nostril vertically to the upper lip and enhances chemoreception (the sensory reception of chemical stimuli). They always have four limbs, four fingers, and five (rarely four) toes. Their coloration differs according to the species; some of the cave-dwelling forms resemble the olm in having a long flesh-colored body and scarlet gill tufts. The smallest are *Thorius* species, some of which reach only 2.7 centimeters (1 inch); the largest is *Pseudoeurycea bellii* at 32.5 centimeters (12½ inches).

As indicated by their common name, these salamanders are always lungless. They breathe through the mucous membrane in the mouth and throat and through the skin, both well supplied with many blood vessels. To keep their skin wet and thus able to absorb oxygen, these animals are linked even more to wet habitats than the lunged salamanders. They shelter in caves, crevices in rocks, spaces between roots and stones, or under logs, and will venture out only when it is humid enough and the temperature is mild. Like most other salamanders they do not tolerate heat well.

Some species, such as cave salamanders, genera *Hydromantes* and *Speleomantes* (a European genus formerly known as *Geotriton* or *Hydromantes*), are mostly rock-dwelling; others, such as *Bolitoglossa,*

are tree-dwellers. Some climbing salamanders (genus *Aneides*) and dusky salamander (genus *Desmognathus*) like to climb on trees or rocks. The shovel-nosed salamander *Leurognathus marmoratus,* and the neotenic and cave-dwelling *Typhlomolge* and *Haideotriton* species are totally aquatic. The many-lined salamander *Stereochilus marginatus* and some *Eurycea, Desmognathus, and Gyrinophilus* species are mostly aquatic.

Each species of lungless salamander has a different type of nuptial dance. Males of species that have a chin gland also have enlarged teeth, with which they scarify the skin of their partner in order to "vaccinate" her with the aphrodisiac secretion of the chin gland.

The lungless salamanders are found almost exclusively in America, where they probably originated, although they also occur in Europe. According to some scientists, the ancestors of the seven European species (all belonging to *Speleomantes,* a genus closely related to the Californian *Hydromantes*) reached western Europe by crossing the Bering land bridge and then became extinct all over Asia and most of Europe; other scientists, including the writers of this chapter, suppose that the ancestors of today's European species had colonized western Europe before its detachment from North America about 50 million years ago.

BENEDETTO LANZA, STEFANO VANNI
& ANNAMARIA NISTRI

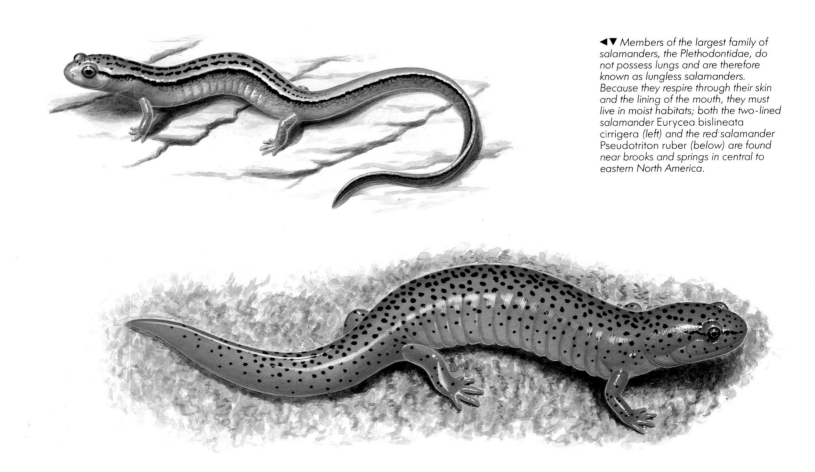

◄▼ Members of the largest family of salamanders, the Plethodontidae, do not possess lungs and are therefore known as lungless salamanders. Because they respire through their skin and the lining of the mouth, they must live in moist habitats; both the two-lined salamander Eurycea bislineata cirrigera (left) and the red salamander Pseudotriton ruber (below) are found near brooks and springs in central to eastern North America.

FROGS & TOADS

KEY FACTS

ORDER ANURA
- 28 families • c. 338 genera
- c. 4,360 species

SMALLEST & LARGEST

Psyllophryne didactyla of Brazil &
Eleutherodactylus iberia of Cuba
Body length: 10 mm (½ in)

Goliath frog *Conraua goliath*
Body length: 30 cm (11¾ in)
Weight: 3.3 kg (7¼ lb)

CONSERVATION WATCH
!!! The 15 species listed as
critically endangered are: golden
toad *Bufo pereglenes*; ferreret
Alytes muletensis; yellow-spotted
treefrog *Litoria flavipunctata*;
armored frog *Litoria lorica*;
Nyakala frog *Litoria nyakalensis*;
peppered treefrog *Litoria piperata*;
Eleutherodactylus karlschmidtii;
Holoaden bradei; *Paratelmatobius
lutzii*; *Thoropa lutzi*; *Thoropa
petropolitana*; gastric brooding
frog *Rheobatrachus silus*; sharp-
nosed torrent frog *Taudactylus
acutirostris*; Mt Glorious torrent
frog *Taudactylus diurnus*; tinkling
frog *Taudactylus rheophilus*.
!! 23 species are listed as
endangered.
! 50 species are listed as
vulnerable.

In Europe, where there are relatively few species of the order Anura, the most conspicuous are smooth-skinned, long-legged frogs of the genus *Rana* and short-legged, warty toads of the genus *Bufo*. So it is only natural that every European language should have specific words for these two kinds. But where species are more diverse, this distinction breaks down, and it is neither possible nor necessary to force animals into one or other category. The name "frog" may properly be used for any member of the Anura, whereas "toad" is loosely applied to members of the genus *Bufo* as well as to other frogs of similar body form.

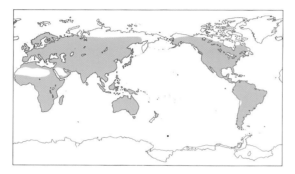

EASILY RECOGNIZED ANIMALS

A frog cannot be mistaken for any other animal. Early in their evolution, frogs acquired a body structure well suited to jumping: the ankle bones are elongated so that, with the femur and tibiafibula, they form a third major segment that gives the hind legs additional mechanical advantage in jumping. The short, rather inflexible vertebral column with no more than ten free vertebrae followed by a bony rod (the coccyx, representing fused tail vertebrae) is another adaptation to leaping. Given such limitations on its structure, a frog—whether it burrows in desert sand, swims in a high Andean lake, or lives in a rainforest tree—is eminently recognizable.

The diversity in body form displayed by frogs expresses their adaptation to particular ways of life. Almost identical solutions to problems of adaptation have evolved time and time again, in different evolutionary lines. Yet any one group of related species may include a diverse range of adaptive types. For example, the so-called "treefrog" family, Hylidae, includes mostly species with the tips of the fingers and toes enlarged into adhesive pads (an aid in climbing) and with the eyes positioned to provide binocular vision ahead, as well as visual fields below and above. But some hylid frogs of semi-arid regions are

▶ *Largest of the North American frogs,
the bullfrog* Rana catesbeiana *is
familiar and widely known by its deep,
bellowing "jug-o-rum!" call. It is
common and widespread, favoring
marshes and wetlands of all kinds,
especially cattail swamps.*

Steve M. Alden/Bruce Coleman Limited

Stephen Dalton/Oxford Scientific Films

adapted for terrestrial life, with eyes set higher on the head, narrow toe pads, and a tubercle on each hind foot that facilitates burrowing. Other hylids may lack any trace of toe pads and live in marshy fields and ditches. These various forms and structures are matched in several other families of frogs. Indeed, given an unfamiliar frog to identify, even an experienced specialist would first have to examine its internal anatomy to ascertain its family relationship, before going on to determine the genus and then the species.

DISTINGUISHING SPECIES
The number of known species of frogs (nearly 4,400) approaches that of mammals (about 4,670). It is impossible to give an exact number because new species of frogs are continually being recognized and given scientific names. This is largely because of discoveries in poorly known regions, especially tropical rainforests. Also, in recent years means other than morphology (form

and structure) have been used to detect differences between species. Many species of frogs prove to be virtually identical in external appearance but can be distinguished by their calls (as indeed they distinguish each other), and with modern electronic techniques for recording and analyzing sounds it becomes possible for biologists to separate these so-called "sibling species". Biochemical techniques, too, make an important contribution by demonstrating genetic (DNA) differences between species whose morphology is virtually the same.

SIGHT, SMELL, AND SOUND
Frogs are primarily visually oriented animals. The eyes are generally large and protrude from their sockets, providing a broad visual field and compensating for lack of rotatory movement. The eyes can, however, be retracted into their sockets, whereupon they bulge against the roof of the mouth and assist in swallowing! The upper eyelid

▲ *Another North American, the leopard frog* Rana pipiens *is extremely well-adapted to cold climates, and its distribution extends northward to Hudson Bay and other parts of northern Canada.*

► The red-eyed treefrog Litoria chloris is common along the coast and hinterland of eastern Australia, roughly from Bundaberg to Sydney. Like many other arboreal frogs, it is slender-bodied and has conspicuously disk-shaped pads on its fingers and toes.

Hans and Judy Beste/AUSCAPE International

is a cover without independent movement. The lower eyelid can be moved and has a transparent upper portion, the nictitating membrane. The pupil is a horizontal oval in most frogs, but in others it is a vertical oval or may even be triangular or diamond-shaped. The iris commonly resembles the side of the head in color and may even be part of some aspect of facial pattern. Presumably this helps to conceal the eye, although some frogs have brightly colored eyes. There is no evidence that frogs have color vision.

A pineal organ (homologous with the "third eye" of reptiles) is found in some frogs and may be detected as a pale spot atop the head. The organ reacts to light, and experiments suggest that it is involved in sun-compass orientation and other aspects of behavior.

Frogs possess both the nasal olfactory organ and a separate vomeronasal organ that functions in a similar fashion. These two organs facilitate homing on breeding sites (presumably by recognition of chemical clues) and may serve in identification of prey.

Hearing is important to frogs, for finding mates and in territorial behavior. The main receptor is a membrane, the tympanum, which is stretched across an oval or round cartilaginous ring behind the eye. A rod of bone, the columella, transmits the vibrations of the tympanum to the inner ear, where sensory cells detect and sort them, passing the information on to the brain. This system is most effective in detecting higher-frequency airborne vibrations, whereas low frequencies pass into the body from the ground or whatever substrate the frog is resting on. The tympanum may be sexually dimorphic, being much larger in males. Frogs of some species have the tympanum concealed beneath skin, but others lack a tympanum completely.

ACTIVITY

Whether frogs are active—that is, engaged in feeding or breeding—depends largely on moisture, temperature, and time of day. Some species are principally nocturnal, others mostly diurnal (day-active), while others may be active day and night. Where moisture is adequate, activity may be limited to times when the temperature is above a level critical for the species concerned. For example, in North America the wood frog *Rana sylvatica* emerges from hibernation and breeds in waters barely above freezing, whereas other species of *Rana* in the same immediate area wait to breed until many weeks later because their eggs are less tolerant to low temperatures. In tropical and subtropical regions,

FROG CALLS

In a word-association test, upon hearing the word "frog" most people would think of "jump" or "croak." How and why do frogs make sounds? Female frogs are for the most part mute. The singing that we hear is done by males. In calling, a frog forces air from its lungs through the larynx, causing the vocal cords to vibrate and produce the sound. The sound is amplified and given a characteristic timbre by the vocal sac or sacs. (Frogs without vocal sacs have very soft voices or may not call at all.) These sacs are pouches of skin beneath the floor of the mouth or at the corners of the mouth, and have openings into the mouth cavity. When calling, a frog keeps its nostrils and mouth closed and uses muscles of the body wall and throat to shunt air back and forth between the mouth-sac cavity and the lungs. The frequency level of the call is largely determined by the frog's size but can be varied somewhat. Other characteristics of the call depend on the pattern of air flow, producing, for example, a long, drawn-out note or perhaps a series of short notes or even clicks. Most calling is done for advertisement: calls serve to attract females ready to mate, and to repel other males from a territory. Some species apparently make one sort of call to serve both purposes, whereas other species have a two-part call: one part of significance to other males, the other to the females. Others have a more varied repertoire, with even an escalating series of territorial calls. One poorly understood kind of call, the so-called "fright scream", is uttered with the mouth open when a frog is seized.

Frog calls are highly species-specific, as we might expect of a behavioral trait that serves to promote selection of an appropriate mate. Listening to a chorus with several species calling simultaneously, a person has no difficulty in distinguishing the different calls. The frogs, however, may hear much less of the sound around them, because their ears tend to be tuned to the frequency level of their own species' call—their hearing is not "jammed" by extraneous noise. Frogs of the same or even different species calling nearby may lessen the confusion by alternating their calls.

C.A. Henley/AUSCAPE International

Scientists, as well as female frogs, make use of calls for distinguishing species of frogs. There are many instances where physical differences between species are very slight or ambiguous but where the calls provide sure identification of the males, at least.

▲ *The call of the bleating treefrog* Litoria dentata *is a penetrating, wavering bleating sound, hence the name. It is found in eastern Australia, especially in paperbark swamps.*

FROGS WITH ALTERNATIVE LIFE HISTORIES

A surprising number of frog species—perhaps 30 percent or more—have a life history differing from the conventional one in which eggs laid in water hatch into tadpoles that grow and metamorphose. Many species omit the free-living tadpole stage and deposit a few relatively large eggs in a moist place out of the water, such as within a rotting log, in a burrow in the ground, or even in epiphytic plants that hold moisture. Usually these eggs are attended by one of the parents, often the male; development takes place within the egg membrane, and the hatchling is a metamorphosed froglet.

Two Chilean frogs, Darwin's frog *Rhinoderma darwini* and the similar *R. rufum*, lay their eggs on land. When the tadpoles hatch, the male picks them up in his mouth and they go into his vocal sac. In the first species the tadpoles develop through metamorphosis in the sac, whereas in the second species the frog carries them to water where they are set loose on their own.

Modifications of life history may involve morphological as well as behavioral adaptations of the parents. The marsupial treefrogs (exemplified by the genus *Gastrotheca* in tropical America) have a brood pouch on the female's back. As the female extrudes the eggs in the course of mating, the male uses his feet to guide them into her pouch. Depending on the species, the eggs may have direct development as described above (with baby frogs being "born" from the pouch) or tadpoles may develop there, later to be released into water to complete their development.

Species of live-bearing frogs occur in Puerto Rico (*Eleutherodactylus jaspari*) and Africa (*Nectophrynoides* and *Nimbaphrynoides* species). The eggs remain in the female's oviducts, where they undergo development through metamorphosis, before the froglets are born.

Michael and Patricia Fogden

Among the many deviations from the standard life history, that of the two species of gastric-brooding frogs of Australia (genus *Rheobatrachus*) is the most bizarre. In these totally aquatic species the female swallows the fertilized eggs, which then develop within her stomach; metamorphosed froglets eventually emerge through the mouth. The stomach's normal digestive mechanism (and, presumably, the urge to eat) is suppressed by a special chemical during this peculiar pregnancy.

▲ Darwin's frog *Rhinoderma darwinii* of South America is notable for its unusual version of parental care. Males gather at clusters of hatching eggs and snap the tadpoles up as they emerge. They are deposited in the male's vocal sacs, where they remain until ready to emerge as small froglets.

Philip Sharpe/Oxford Scientific Films

◄ Another South American, the marsupial frog *Gastrotheca ovifera* has another parental care strategy in which the female "incubates" the eggs in a pouch on her back.

where low temperatures are less likely to be important, active periods nevertheless may be controlled by seasonal changes in moisture. This is especially true for species that live away from permanent sources of water and depend on rainy seasons for both feeding and breeding activity.

BREEDING

With some specialized exceptions, frogs fertilize their eggs externally. The male approaches the female from behind and grasps her around the body with his arms. The hold may be at the waist (considered the primitive mode) but in most species is just behind the female's arms or (rarely) around the head. Males of many species have horny areas on their hands that help in gripping a slippery mate. The frogs maintain this posture, while the eggs are fertilized as they are extruded. When eggs make contact with water, the jelly membranes surrounding them swell, encasing each egg in a transparent sphere.

Eggs deposited in water are most commonly grouped in globular masses containing a few to hundreds of eggs. Some species that breed in still, warm water likely to be depleted of oxygen spread the eggs in a single layer on the surface, whereas those breeding in fast-flowing streams may attach the eggs singly to submerged rocks. Eggs deposited out of water may be in a variety of moist locations such as in tree holes, in plants growing on trees, underneath moist leaf-litter, or in holes in the ground. These are only a few examples of the many known sites and modes of egg deposition.

LARVAL LIFE

Frogs resemble other amphibians and differ from reptiles in that most species (75 to 80 percent of frogs) have a larval stage interposed between the egg and the mature body form. This period may be as brief as one week in species breeding in short-lived desert rainpools, or up to two years or more, but generally it averages a few weeks. A typical frog larva, or tadpole, lives in water and has a rather oval body with a strong finned tail, but no clear distinction between head and body. The mouth has a beak and rows of "teeth" made of a chitinous substance. Water taken in through

Michael Fogden/Oxford Scientific Films

▲ *Centrolenella valerioi is typical of most of the arboreal glass frogs of Central and South America. A female lays her eggs on the underside of leaves overhanging running water, and the male then guards them until they hatch and fall, as tadpoles, into the water below.*

Patrick Clement/Bruce Coleman Limited

◄ *Fertilization is external in almost all frogs, and the male typically clasps the female in a tight embrace, known as amplexus. This embrace may be maintained for several days, perhaps as the least complicated means of ensuring he is on the spot when she is ready to lay her eggs. These are European common frogs Rana temporaria, which lay their eggs in gelatinous masses.*

Bach/ZEFA

▲ A golden mantella *Mantella aurantiaca* draws a bead on an unsuspecting fly. In most frogs and toads the tongue is attached at the front, free at the back, and can be flicked forward some distance and with considerable speed . . .

Michael and Patricia Fogden

► Generally speaking, adult frogs are carnivores but their larvae are vegetarians. Nevertheless, there are some predatory tadpoles, and cannibalism is not unusual. Even vegetarian larvae, such as these meadow treefrog tadpoles *Hyla pseudopuma* may occasionally make a meal of a dead sibling.

the mouth passes over gills concealed within a chamber behind the mouth, before being expelled through the spiracle, a hole usually on the left side of the body. The gills not only serve for respiration, but also filter tiny food particles from the water, diverting them toward the stomach and supplementing the algae and detritus obtained by biting and scraping. Also concealed within the gill chamber are the animal's growing front legs. The rear legs develop externally where the tail and body meet. At the end of the larval stage the tadpole undergoes metamorphosis into the adult form. This transformation is more than simply growing limbs and resorbing the tail. Profound changes take place in both morphology and physiology. For example, the digestive tract shortens, becoming suited to a carnivorous diet

Boch/ZEFA

instead of a vegetarian diet; the gills disappear, and respiration is taken over by the lungs (tadpoles also may use their lungs long before metamorphosis is completed); the skeleton is extensively modified, and true teeth develop in the upper jaw.

FOOD AND ENEMIES

All frogs are carnivores, and many are generalized feeders that will take whatever small animals— vertebrate or invertebrate—their capacious mouths can accommodate. Relatively few frogs are large enough to eat other vertebrates, so most of them eat insects and other arthropods, and earthworms. But a large frog such as the North American bullfrog *Rana catesbeiana* can take birds and mice, small turtles and fish, and under

crowded conditions is a fearsome predator on smaller frogs of its own and other species. Tadpoles for the most part are vegetarians— filtering organisms from the water, scraping algae from stones, consuming bottom debris. Some species have predaceous tadpoles, however, which capture invertebrates or other tadpoles.

Frogs are prey to a host of enemies, ranging from tarantulas to humans. Many snakes live largely or entirely on frogs. Herons skewer frogs in shallow water, bats snatch them from branches over tropical pools, turtles ambush them from under water, big frogs eat smaller frogs, parasitic flies lay eggs on them, and leeches attack them externally and even enter their bodies. The larval stage, too, is a perilous time. Almost any meat-eater finds a tadpole a succulent morsel.

. . . snapping up prey with the aid of sticky secretions from glands in the tongue. Vision seems to be an important hunting sense, although there is no indication that frogs have color vision.

► One of a small group of primitive frogs restricted to New Zealand, Archey's frog Leiopelma archeyi lives on rough mountainsides. The eggs are laid on damp ground under logs or rocks, and the young are carried about on the male's back until they reach independence.

Frances Furlong/Bruce Coleman Limited

WHERE FROGS LIVE

Frogs are native to all the continents except Antarctica but are absent from almost all remote oceanic islands and from Greenland and other Arctic islands. Although widely distributed over the Earth, frogs are far more abundant and diverse in moist areas of the tropics, especially tropical rainforests, than farther north or south; regions closer to the poles are subject to extremes of weather, and dryness and cold are especially hostile to frog life.

The near-absence of native frogs from oceanic islands is rooted in the geological history of the islands, most of which are volcanoes or coral islands built on volcanic footings. There has simply been no way for animals so sensitive to desiccation and heat to reach them (frogs cannot live in salt water). Many of these islands may have suitable habitats; for example, several species of frogs have been introduced to the Hawaiian islands, where they prosper. Some islands—such as the Fiji group in the Pacific and the Seychelles group in the Indian Ocean—are unusual in having endemic frogs (species found nowhere else), presumably because these landmasses were formerly part of or close to continents but were moved by geological processes, carrying with them the ancestors of the present frog species.

ANCESTORS AND RELATIVES

All living amphibians—frogs (order Anura), salamanders and newts (order Caudata), and caecilians (order Gymnophiona)—are thought to be more closely related to each other (in the subclass Lissamphibia) than to other groups of amphibians now extinct but recognized in the fossil record. The relationships of the Lissamphibia are not well understood, but their

common ancestry may be with the Dissorophoidea amphibians of the Palaeozoic era, more than 245 million years ago. Whatever their exact ancestry, today's amphibians lie on an ancient branch of the amphibian tree and are not on the evolutionary line that led to reptiles and, eventually, mammals.

The oldest frog-like fossil is *Triadobatrachus* found in Madagascar in deposits of the Triassic period, more than 200 million years ago. Early frog fossils are also found in deposits laid down early in the Jurassic period, almost 200 million years ago, in Argentina. By the beginning of the Tertiary period 65 million years ago, there were frogs we can classify in genera that are still represented today. So frogs that differed in no essential respect from those alive today were contemporaries of the dinosaurs, survived the great extinctions at the close of the Mesozoic, and persisted largely unchanged while mammals underwent their great evolutionary expansion.

ARCHAEOBATRACHIAN FROGS

There are four suborders of modern frogs. Some are considered more "primitive" (in evolutionary terms) than others.

The suborder Archaeobatrachia includes four families, Ascaphidae, Bombinatoridae, Leiopelmatidae, and Discoglossidae, thought to be the most primitive living frogs. A clearly primitive character they share is the presence of free ribs (all other frogs have their ribs fused to the vertebrae).

Ascaphid and leiopelmatid frogs

There are only four species of Ascaphidae and Leiopelmatidae alive today. The tailed frog *Ascaphus truei* lives in cold, fast-flowing mountain streams of the northwestern United States and southwestern

Canada. The so-called "tail", is present only in the male and is an organ adapted for insertion, so that the eggs are fertilized internally before being released from the female's body (in this type of habitat, external fertilization might be unsuccessful) and attached to the undersides of rocks in the water. The sucker-mouthed tadpoles take three years to develop and are well adapted to life in fast-running waters.

The other three species are restricted to New Zealand and are the only native frogs there. One of these, Hamilton's frog *Leiopelma hamiltoni*, occurs on just two small islands and is at risk because of its tiny area of distribution. They are all small frogs, no bigger than 50 millimeters (2 inches) body length. *L. hamiltoni* and Archey's frog *L. archeyi* lay their eggs on moist ground under rocks, logs, or vegetation; the hatchlings are in a relatively late state of larval development and climb upon the back of the attending adult male. Hochstetter's frog *L. hochstetteri* lays eggs in shallow hollows with seepage; the tadpoles are more adapted for swimming and remain in the water near to where they hatched until metamorphosis, although not feeding.

Subfossil remains, which date back to the period before humans colonized New Zealand, reveal that there were additional species of *Leiopelma*, one twice the size of the living species. Fossils from early and late in the Jurassic period (up to almost 200 million years ago) found in southern Argentina, are considered to belong to the family Leiopelmatidae.

Bombinatorids and discoglossids

The bombinatorid and discoglossid frogs are only slightly more numerous but are less geographically restricted than the ascaphid and leiopelmatid frogs. All but two of the 16 species live in the Northern Hemisphere north of the tropics. Species of the genus *Discoglossus* are semi-aquatic frogs found in Europe, North Africa, and the Middle East. Different species of fire-bellied toads (genus *Bombina*) are small to moderate-sized frogs—up to about 70 millimeters (almost 3 inches) body length in a large Asian species—that live in the Far East and Europe, where they have a mostly aquatic existence in shallow water habitats. When disturbed they arch their back and throw up their arms and legs,

▼ *The tailed frog* Ascaphus truei *(below left) belongs to the family Ascaphidae, whose sole member is found in western North America only. Although they possess primitive tail-wagging muscles, none has a true tail. The European painted frog* Discoglossus pictus *(bottom left) belongs to the family Discoglossidae and possesses a number of primitive features such as a non-projectile, disk-shaped tongue and round pupils. A member of the Bombinatoridae, the oriental fire-bellied toad* Bombina orientalis *(below right), displays its brightly colored underside, warning of a mildly toxic, unpalatable skin secretion.*

▲ *Mute, tongueless, and built something like a squared-off pancake with a limb at each corner, the Surinam toad Pipa pipa is one of the more bizarre members of the tropical South American frog fauna. It is almost entirely aquatic.*

showing the bright red or yellow colors beneath. Because the frog's skin secretions are distasteful, this is considered a form of warning behavior. The midwife toads (genus *Alytes*) are largely terrestrial frogs of Europe and North Africa, known for the male's care of the eggs; these adhere to his back and thighs and are carried about in and out of water until hatching time, when he enters the water, allowing the tadpoles to swim away. The two species of *Barbourula* are exceptional among discoglossids, not only in their geography—one in the southern Philippines, the other south of the Equator in Borneo—but also in being wholly aquatic. Their adaptation to life in streams includes having broadly webbed fingers and toes. The fossil record shows discoglossids to have existed in Europe as far back as the Jurassic, about 150 million years ago, and the family was present in North America in the late Cretaceous period, more than 65 million years ago.

PIPOID FROGS

Members of the suborder Pipoidea, with two families, the Pipidae and the Rhinophrynidae, are among the most peculiar of frogs. The pipids

(30 species, 5 genera) have flattened bodies, are tongueless (all other frogs have tongues), and except for frogs in the genus *Pipa* have claw-like structures on three of the toes. All species are highly aquatic, with large, fully webbed feet. They don't have vocal cords but produce clicking sounds under water by using bony rods in the larynx. The size range is from about 40 to 180 millimeters (1½ to 7 inches) body length. Frogs in the genus *Pipa* live in South America east of the Andes and in eastern Panama. The Surinam toad *Pipa pipa,* which is almost 180 millimeters (7 inches), is bizarre in both appearance and breeding habits. The eggs becomes embedded in pockets in the skin of the mother's back, where they develop to emerge as tiny frogs. In another *Pipa* species, tadpoles emerge from the chambers and complete their development on their own.

In Africa south of the Sahara Desert there are three genera. The African clawed frogs, genus *Xenopus,* were for a time medically important as animals used in testing for human pregnancy: urine from a pregnant woman injected into a female frog causes it to extrude eggs. Development of chemical tests has spared platannas (as they

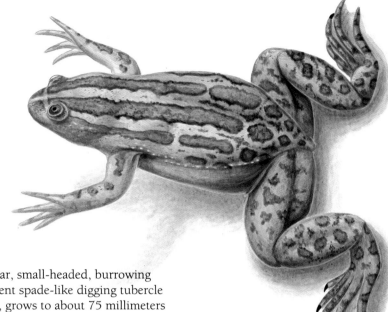

are called in South Africa) from this indignity, although they continue to be important animals for laboratory research because of the ease with which they may be kept and bred in captivity.

Fossils of pipid frogs from the early Cretaceous period, perhaps 145 million years ago, have been found in Israel, and from the late Cretaceous, more than 65 million years ago, in Africa and South America. Clearly, their distribution dates from a time when the southern continents were part of Gondwana, before Africa and South America became separate landmasses.

Some scientists classify the Mexican burrowing toad *Rhinophrynus dorsalis* with the pipid frogs; others place it in its own suborder (Rhinophrynoidea). It is the only living representative of the family Rhinophrynidae.

This rather globular, small-headed, burrowing frog has a prominent spade-like digging tubercle on each hind foot, grows to about 75 millimeters (3 inches), and ranges from southern Texas to Costa Rica. It emerges after heavy rains to call and breed in temporary ponds. Fossils show that in the late Paleocene to the early Oligocene epochs (about 37 million years ago) related species lived as far north as southern Canada.

PELOBATOID FROGS
The suborder Pelobatoidea comprises three families, Megophryidae, Pelobatidae, and Pelodytidae, although some scientists treat them as one family. These frogs are considered rather primitive, but in some ways are transitional to more advanced frogs, and have in common the supposedly primitive

▲ The Mexican burrowing toad Rhinophrynus dorsalis *(left) is not a true toad, but has evolved a number of toad-like features. It is usually seen above ground only following heavy rain, when it emerges to breed. Rarest of the clawed frogs of Africa, the Cape clawed frog* Xenopus gilli *(right) belongs to an ancient family of aquatic frogs, the Pipidae. The hind feet have black horny caps on three of the toes.*

Chris Mattison

◄ The Surinam toad's breeding habits are nearly as bizarre as its appearance. In a complicated underwater ritual, the male fertilizes his mate's eggs as they are laid, and distributes them over her back, where they sink into the spongy tissue over several days. The eggs hatch into tadpoles in the pockets formed, then continue their development until they leave as small froglets.

▲ Spadefoots, like this Couch's spadefoot Scaphiopus couchii (left) possess a spade-like tubercle on the bottom of their hind feet which they use to burrow rapidly backwards, circling as they descend. The Asian horned toad Megophrys nasuta (right) is adapted for life on the rainforest floor, where its shape and "dead-leaf" coloration render it almost invisible among the leaf-litter.

habit of the mating male clasping the female around the waist or groin, rather than higher on the back, just behind the arms. The eyes of all species have vertical pupils, and all have free-living, aquatic tadpoles.

Megophryids, pelobatids, and pelodytids

The family Megophryidae has fewer than 90 species recognized in nine genera found in Asia, from Pakistan and western China to the Philippines and through Indonesia to the Greater Sunda Islands in the Malay Archipelago. Their habitats vary from forested tropical lowlands and uplands to Himalayan peaks. *Megophrys* species inhabit the forest floor; they have funnel-mouth tadpoles which develop in still waters. *Megophrys montana* is notable for its cryptic coloration and form; not only does it resemble dead leaves in color and pattern, but one subspecies has pointed projections from the eyelids and nose that disrupt the frog-like outline. Also inhabiting the forest floor are *Leptobrachium* species; these frogs breed in small streams, and the tadpoles live amid rocks and gravel. Mountains as high as 5,200 meters (17,000 feet) are the home of *Scutiger* species, which lead a largely terrestrial existence but breed in cold mountain streams. An odd feature of the *Scutiger* male is that his mating grasp of the female is enhanced by two spiny patches on the chest in addition to the more usual spiny areas on the first two fingers.

"Spadefoot toads" is the common name for the three genera in the family Pelobatidae: *Pelobates,* found in Europe, some areas of Mediterranean northwestern Africa, and western Asia; and *Scaphiopus* and *Spea,* found in North America from southern Canada to southern Mexico. They are toad-like burrowing frogs of moderate size, up to 100 millimeters (4 inches) body length, and their common name comes from

the prominent digging tubercle on each hind foot. The four species of *Pelobates* are partial to areas of sandy soil, where they burrow and also breed, in pools (often temporary ones) and ditches. *Scaphiopus* and *Spea* too spend their inactive periods in self-made burrows; of the seven species, one lives in humid regions of the eastern United States, the others in arid areas, even deserts. Conditions in humid areas may be adequate for surface activity during the warm period of the year, but where rainfall is scarce the frogs are forced to spend most of the year underground. When the soil temperature is warm enough, the sound of rain falling will prompt them to emerge, and they migrate to temporary pools of water to breed. Embryonic and larval development is rapid—less than two weeks may elapse between egg-laying and metamorphosis. But even this may not be rapid enough if additional rain doesn't fall to offset evaporation in the fierce desert heat. Some spadefoot tadpoles develop a different morphology—enlarged jaw muscles and a large beak—from others of their species, and they cannibalize their pondmates, thus making the most of the available protein. The rains that call out the spadefoots also bring forth hordes of termites and other insects, and the frogs can get enough food in a few nights to survive many more months underground.

Pelobatid fossils occur as early as the late Cretaceous of North America and Asia (more than 65 million years ago), and the modern genera *Scaphiopus* and *Pelobates* appear in the Tertiary of North America and Europe, respectively, 35 to 47 million years ago.

The family Pelodytidae has two living species in the single genus, *Pelodytes;* one in Europe; and one further east, in the Caucasus region (between the Black Sea and the Caspian Sea). These terrestrial frogs look like small (50 millimeters, or 2 inches) more-typical frogs, being long-legged and lacking the digging spade. The earliest fossils are from the middle Eocene epoch in Europe (47 million years ago). From North America, where the family is extinct, there is a middle Miocene fossil some 15 million years old.

THE LAST SUBORDER: NEOBATRACHIA
The suborder Neobatrachia, thought to represent the more advanced frogs, includes more than 4,200 species, or about 96 percent of all living species of frogs. There are 17 families.

Leptodactylid frogs of the Americas
The largest family in this suborder is the Leptodactylidae, a diverse assortment of almost 1,000 species classified in 48 genera, inhabiting South America, the Caribbean islands, Central America, and Mexico; only six species cross into the southern United States. The world's southernmost frog is a leptodactylid, *Pleurodema bufonina,* which reaches the Strait of Magellan.

Within the vast array of leptodactylids are virtually all modes of life open to frogs. In the genus *Eleutherodactylus,* which has over 600 species, many spend their active hours in trees or shrubs, but others forage on the leaf-litter of the forest floor, and all but one *Eleutherodactylus* have direct-developing eggs (the exception is viviparous). Species of several other genera have rather cryptic lives on or within forest leaf-litter; others that live in arid regions remain underground, encased in a protective "cocoon", until rainfall permits a brief emergence for feeding and breeding; and many species have more familiar habits in and around streams and ponds. Wholly aquatic species inhabit lakes in the Andes.

Morphology and lifestyle are related, so the variations in their body form are a reflection of their ecological diversity. For example, tree-dwelling *Eleutherodactylus* are rather small, slender frogs with features found in other treefrogs, such as enlarged disks on fingers and toes, and large forward-directed eyes. Burrowing species of other genera are short-legged, with a prominent spade on each hind foot for digging.

Michael and Patricia Fogden

Leptodactylid frogs of several genera lay their eggs in foam "nests"; according to the species involved, such a nest may float on open water, be in a cavity adjacent to water, or be in a burrow or other sheltered site on land. The nest is constructed while the frogs mate, the male using his feet to whip the mixture of eggs and seminal

▲ *Eleutherodactylus is a very large genus of frogs inhabiting Latin America. In all but one of the more than 600 species, metamorphosis takes place entirely in the egg, the young finally emerging as fully formed small froglets.*

◄▼ *The western barking frog* Eleutherodactylus augusti cactorum *(left) inhabits limestone caves and crevices in southwestern North America and adjacent Mexico. The ornate horned toad or escuerzo* Ceratophrys ornata *(below left) is a robust, aggressive predator of the rainforest floor. Widely eaten by humans, the South American bullfrog* Leptodactylus pentadactylus *(bottom right) is one of the largest members of the family Leptodactylidae.*

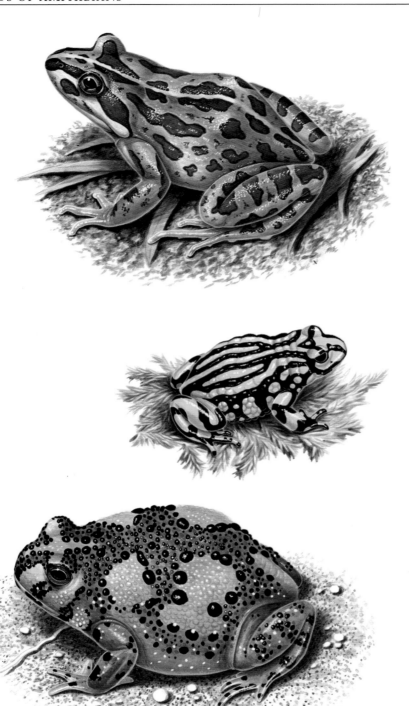

▲ *Three members of the Australasian family Myobatrachidae: the spotted grass frog Limnodynastes tasmaniensis (top) inhabits semi-aquatic niches; the corroboree frog Pseudophryne corroboree (center) lives in the Australian Alps and is so named because its vivid coloration is similar to the striped body decorations adopted by the Aborigines for ceremonial dances, or corroborees; and the crucifix toad Notaden bennettii (bottom), which has a specialized diet of ants and termites.*

the female *Leptodactylus ocellatus* remains with the nest, and when the tadpoles hatch they stay together in a school and the female protects them against predators.

The presence of leptodactylids in South America throughout the Cenozoic (beginning 65 million years ago) is documented in the fossil record. A most remarkable fossil from the Dominican Republic on the Caribbean Island of Hispaniola is an *Eleutherodactylus* preserved whole in amber, at 37 million years old.

Myobatrachid frogs of Australasia

The myobatrachid frogs of Australia and New Guinea, numbering about 120 species in 23 genera, are the eastern hemisphere counterparts of the leptodactylids of North and South America. Indeed, some people classify them as members of the Leptodactylidae. Australia has a large proportion of arid land, where rainfall is sparse and seasonal, so it is not astonishing that many of the myobatrachids are burrowers and are seldom seen or heard except when they emerge to mate and deposit eggs in temporary pools. The more conventional burrowers such as members of the genera *Heleioporus, Notaden,* and *Neobatrachus* are squat, short-legged frogs with a digging spade on each hind foot. The turtle frog *Myobatrachus gouldii* and the sandhill frog *Arenophryne rotunda,* both of southwestern Australia, are exceptional in that they deposit their eggs deep in moist sand, as deep as 1 meter (3¼ feet), where they undergo direct development without a free-living tadpole stage. Both are also unusual in that they burrow head-first (rather than rump-first as most burrowing species do) into the sand or soil.

In contrast to the Leptodactylidae, the Myobatrachidae family does not include any treefrogs (this niche in Australia and New Guinea is reserved for members of the Hylidae and Microhylidae), but still there is much diversity in addition to the burrowers, especially in the tropical and temperate rainforests of eastern Australia. Examples include small frogs of the genus *Taudactylus,* at home in torrential creeks of the northeastern rainforests; and tiny *Ranidella,* about 25 millimeters (1 inch) in length, which are abundant in most of Australia's moist habitats. The latter pose a problem for herpetologists because of their morphological similarity—some species can be distinguished only by differences in their calls.

Two species of *Rheobatrachus,* the gastric brooding frogs, are highly aquatic stream-dwellers that scarcely became known to science before they disappeared for no obvious reason, possibly having become extinct. Another species with a less strange but nevertheless odd form of parental care is the pouched frog *Assa darlingtoni,* a ground-dweller of forests on the Queensland–New South Wales border. The male of this secretive

fluid (and water, if present) into a froth. The nest protects the fertilized eggs and then the tadpoles from enemies, from overheating (if exposed to the sun), and from desiccation. A nest in a cavity can shelter the tadpoles until rains flood the site and the tadpoles escape to continue their development, as occurs in some species of *Leptodactylus.* In species of *Adenomera,* which create their nests in moist cavities on land, the tadpoles remain in the nest without feeding until they metamorphose. Hatchlings from nests on open water may merely swim off, but some surprising exceptions are known. For example, species has a brood pouch on each side of the body. When the eggs (which are laid on land) hatch, the male straddles the mass and the tadpoles make their way into the pouches. Further development and metamorphosis take place there, and the froglets emerge some seven to ten weeks later.

Frogs in several Australian genera make a foam nest for the eggs, but unlike the American leptodactylids—where the male's feet do the job—in Australia the mating female beats the foam with her hands, assisted by flanges on some of the fingers. Some create their foam nests on open water, others in water-filled burrows, and several create them on land. At least four recent genera of myobatrachid frogs (*Lechriodus, Limnodynastes, Crinia,* and *Kyarranus*) were present in Australia in the mid-Tertiary, 15 to 25 million years ago.

Toads and harlequin frogs

Toads (family Bufonidae) are native to temperate and tropical zones, deserts and rainforest, mountains and prairies, everywhere except the Australian region, Madagascar, and oceanic islands. Of about 400 species in the family, more than half belong to one of the 33 genera, *Bufo*. Toads of this genus range in size from 25 millimeters (1 inch) body length for some African

◄ Newly hatched larvae of the pouched frog Assa darlingtoni of eastern Australia in the act of entering the male's pouches. The male assists the young by gentle scooping movements of his arms and legs and usually ends up with four or five in each pouch; any left over die within a few hours.

H. Ehmann

species to 25 centimeters (10 inches) for giant toads *Bufo blombergi* and *B. marinus* of the wet forests of South America—large enough to fill a dinner plate.

No matter what their size, all *Bufo* species conform to a standard appearance: heavy-set, short-legged, with numerous wart-like glands on the body and legs, and a prominent, rounded or elongate parotoid gland behind the eye. Also, they are toothless, although this is evident only on close examination. Habits tend to be rather

◄▼ The strikingly marked leopard toad Bufo pardalis (left) of southern Africa is a typical true toad, with short limbs, dry warty skin, and large parotoid glands. The Asiatic climbing toad Pedostibes hosii (bottom left) shows characteristics of a tree-dwelling existence, such as long slender limbs and broad adhesive pads on the digits. Looking more like the unrelated poison frogs than its relatives the toads, the variable harlequin frog Atelopus varius (below) of Central America is brilliantly colored, a warning of its toxicity.

▲ *The golden toad* Bufo periglenes *is known only from the rainforests of Costa Rica's Monteverde Cloud Forest Reserve. Except for the few days in each year on which they emerge and congregate to mate, they spend their lives in cavities amid the root systems of forest trees. The last emergence was in 1987, and subsequent searches have failed to find it; the species may now be extinct.*

similar among the species, too. They are ground-dwellers, typically hiding in holes during the day and emerging at night to hop about looking for invertebrates, which they snap up with their long sticky tongue.

Breeding is remarkable only for the immense numbers of eggs that many species produce. The eggs typically are laid in paired strings (one from each ovary) and may number 20,000 or more from a toad only 70 or 80 millimeters (about 3 inches) long. All frogs have a variety of skin glands, and in many species they produce defensive secretions, bad-tasting or even deadly poisonous. Among the most effective are toad poisons, concentrated especially in the parotoid glands; a dog that mouths a giant Colorado River toad *Bufo alvarius*, for example, may be fatally poisoned. This does not mean that you'll be poisoned if you handle a toad (or get warts, for that matter!), but it is a good idea not to rub your eyes after handling any frog, until you've washed your hands.

Atelopus is a Central and South American genus (more than 60 species) of small, often brightly colored, slow-moving toads, whose sucker-mouthed tadpoles live in rapidly flowing water. These diurnal toads produce potent toxins, so the conspicuous color patterns may serve to warn potential predators. Some of the remaining genera—all of which have fewer than ten species—are much like *Bufo* in structure and habits, but others diverge. For example, *Nectophrynoides* of Africa is noted for giving birth

to metamorphosed froglets; *Pedostibes* of Southeast Asia is a tree-dweller; and *Ansonia,* also of Southeast Asia, lives on the ground but breeds in riffles or cascades of water, and its tadpoles are adapted to torrents.

Bufonid fossils occur in the upper Paleocene of South America (more than 57 million years ago) and in the later Tertiary of North America, Europe, Asia, and Africa.

Hylid treefrogs
The treefrog family, Hylidae, has more than 770 species in 40 genera. More than 500 of these species live in the Americas, especially the tropics. There are more than 140 species in the Australia–New Guinea area; and a mere 12 closely related species of the genus *Hyla* are distributed across temperate regions of Eurasia, scattered from Spain to Japan, one of which also occurs in northwestern Africa.

Most species in the family are arboreal or at least climbing forms, which show the characteristic adaptations to this way of life: fingers and toes with expanded tips having adhesive properties, and eyes placed somewhat laterally and forward-directed, enhancing vision downward and binocular perspective. (Terrestrial and aquatic frogs generally have the eyes more atop the head and aimed less forward, thus giving them a larger horizontal visual field and permitting aquatic species to rest with only their eyes above water.) Treefrogs of the genus *Phyllomedusa* in tropical America have the first finger opposable to the other three, like a thumb, permitting them to grasp twigs and stems. Like other large families, the hylids have undergone adaptive radiation into several major ecological niches. There are no thoroughly aquatic species, but a number of terrestrial species are typically associated with ponds and marshes, and in wet weather they may range more widely. These are long-legged frogs, powerful jumpers with webbed toes; the tips of their fingers and toes are pointed or only slightly expanded, in contrast to those of their climbing relatives. Some of these occur in the same genus as that of tree-dwelling species—the large Australasian genus *Litoria*, for example, which also includes frogs adapted to habitats other than solely arboreal or solely aquatic. There are burrowing "treefrogs" too. Members of the Australian genus *Cyclorana* spend much of their lives underground, and one species is noted for the large amount of water it can store to tide it over periods of drought. These, and the North American *Pternohyla*, form a "cocoon" while underground for protection against desiccation.

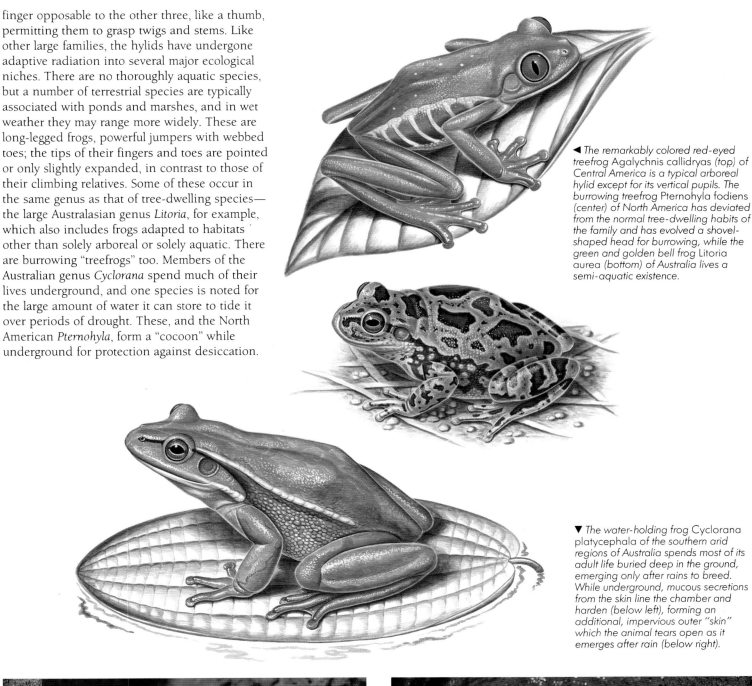

◀ The remarkably colored red-eyed treefrog Agalychnis callidryas (top) of Central America is a typical arboreal hylid except for its vertical pupils. The burrowing treefrog Pternohyla fodiens (center) of North America has deviated from the normal tree-dwelling habits of the family and has evolved a shovel-shaped head for burrowing, while the green and golden bell frog Litoria aurea (bottom) of Australia lives a semi-aquatic existence.

▼ The water-holding frog Cyclorana platycephala of the southern arid regions of Australia spends most of its adult life buried deep in the ground, emerging only after rains to breed. While underground, mucous secretions from the skin line the chamber and harden (below left), forming an additional, impervious outer "skin" which the animal tears open as it emerges after rain (below right).

Michael Fogden/Oxford Scientific Films

▲ *Adults of many poisonous frogs have conspicuous warning colors, but the phenomenon is less widespread among tadpoles. But there are exceptions, including perhaps this colorful tadpole of the Central American treefrog* Hyla ebraccata.

The breeding habits of hylids are, with some notable exceptions, fairly conservative, and most species have aquatic tadpoles. Eggs of torrent-dwelling species are fixed firmly to rocks. Those in still water may be attached to aquatic vegetation, or spread in a thin film on the surface, thus ensuring adequate oxygen in warm waters suffering from oxygen depletion. A common tendency in regions of high humidity is to place the eggs on vegetation emerging from the water or even on leaves high in trees overhanging the water, so that when the tadpoles hatch they drop into the water. Some forest species breed in water-filled tree holes, and others in water-holding epiphytic plants such as bromeliads.

Tadpoles of many hylid species are midwater pond-dwellers, with laterally placed eyes (the better to see below as well as above) and a broad-finned tail tapering to a filamentous tip. These tadpoles may be seen hanging at an angle in the water with the tail tip vibrating while they filter microscopic food particles. Stream- and torrent-dwelling tadpoles, such as those of *Nyctimystes* in New Guinea, are quite different: depressed body, low tail fins, dorsal eyes, and mouthparts formed into an oval sucker.

Several tropical American genera (all classified in the subfamily Hemiphractinae) have the peculiar habit of brooding the eggs on the female's back: in the genus *Hemiphractus*, the eggs are exposed on her back; in the genus *Gastrotheca,* they are completely enclosed within a pouch; and there's a whole range of body-behavior adaptations in between. One species of this subfamily deserves special mention apart from its breeding habits: *Amphignathodon guentheri* is the only living species of frog with teeth in its lower jaw. Other frogs may be toothless or possess teeth on the upper jaw, or on the upper jaw and the roof of the mouth, and some have large fang-like bony projections at the anterior ends of the lower jaws, but this species is unique in having true teeth.

Fossil hylid frogs are known from the Paleocene of South America (more than 57 million years ago) and the middle to later Tertiary of North America, Europe, and Australia.

Glass frogs

The glass frogs, family Centrolenidae, are a group of more than 120 species, most of them tree-dwellers, inhabiting moist forests from southern Mexico to Bolivia, plus southeastern Brazil-northeastern Argentina. The "glass frogs" name derives from a scarcity of pigment in the skin of the abdomen, which makes the internal organs visible. Most of the species are small (maximum size about 30 millimeters, or 1¼ inches body length) and green. Two larger species attain 75 millimeters (3 inches). Three

◄ *Glass frogs, such as* Hylinobatrachium fleischmanni, *are small tree-dwellers, inhabiting moist forests from southern Mexico to Bolivia, and southeastern Brazil and Argentina.*

genera are recognized: *Centrolene, Cochranella,* and *Hylinobatrachium.*

Males of most centrolenids call from leaves overhanging streams and then remain near the eggs laid on leaves at these calling sites. Egg masses placed in such situations are safe from many predators but are parasitized by flies which lay their eggs on the mass so that the maggots consume the frog eggs. Frog larvae that survive the parasites fall to the stream, where they live in gravel or debris. These tadpoles are elongate with muscular tails and very low fins (broad fins are useful only to tadpoles that swim in open water).

Some larger centrolenids live and breed in rocky waterfalls, where the egg masses are stuck to rock surfaces. No fossil centrolenid is known.

Dendrobatids

The family Dendrobatidae has some of the most colorful and interesting frogs. Most are rather small, the smallest less than 15 millimeters (½ inch) body length, although one species reaches 62 millimeters (almost 2½ inches). There are more than 170 species in eight genera. They inhabit moist tropical regions in Central and South America, from Nicaragua to southeastern Brazil

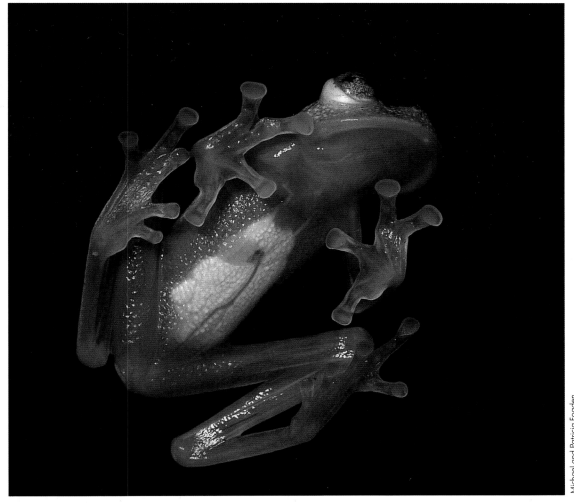

◄ *The bare-hearted glass-frog* Hylinobatrachium colymbiphyllum *and its relatives are so called from the translucent skin of the underparts, through which the heart and other internal organs can easily be seen.*

Michael and Patricia Fogden

▲ Many poison frogs of the genus *Dendrobates* of Central and South America display the kind of brilliant colors and bold patterns normally associated only with butterflies, hummingbirds, and coral-reef fishes. The skin secretions of these small frogs are among the most toxic substances known, and are used by forest Amerindians to poison their blow-gun darts. This species is *Dendrobates tinctorius*.

and Bolivia. Unlike the majority of frogs, almost all dendrobatids are diurnal (active during the day). They lay small numbers of eggs in moist sheltered places, and a parent (usually the male) guards the eggs. When the tadpoles hatch they wriggle onto the parent's back, are carried to water, and released to complete their development. Some species have a peculiar variant of this behavior: the female releases the tadpoles into a water-holding bromeliad plant, and she returns occasionally to deposit an unfertilized egg as food for the tadpoles.

The dendrobatids fall into two main groups. One includes a large number of mostly dull-colored species of the genera *Colostethus, Mannophryne,*

and *Nephelobates* which live alongside streams or on the forest floor and, with one known exception, are non-toxic. The second group consists of the poison frogs (genera *Dendrobates, Phyllobates, Epipedobates,* and *Minyobates*) which are very colorful and whose skin glands excrete alkaloid poisons that act on the nervous system. In addition there is the genus *Aromobates,* with a single species *A. nocturnus,* unique on several counts: it is the largest dendrobatid, it gives off a foul, presumably protective odor but is not poisonous, and it is a nocturnal stream-dweller.

The toxicity of the poisonous dendrobatids varies greatly from species to species. Presumably the poisons are a defense against predators, and the bright color patterns act as a warning. One species deserves special mention: *Phyllobates terribilis* is so poisonous that it is unsafe even to handle. The toxins in one frog's skin would be sufficient to kill more than 20,000 laboratory mice, and less than 200 micrograms introduced into a human's bloodstream could be fatal. The toxicity of this species and others less-poisonous has long been known to certain groups of Indians of western Colombia, who use the frogs to poison their blow-gun darts (not arrows, as the popular literature often states). The toxin of *P. terribilis* is so abundant and potent that merely rubbing the point of a dart across a living frog's back is sufficient to make it deadly in hunting.

No fossils of dendrobatid frogs are known.

Ranids or "true" frogs

The "true" frogs, family Ranidae, have the widest distribution of any frog family: North America (even in Alaska), Central America, and northern South America; Europe and across Asia south of the Arctic Circle, through the East Indies to New Guinea, the extreme north of Australia, and the Fiji islands; and most of Africa, and Madagascar.

Jany Sauvanet/AUSCAPE International

▼ The poison frogs of the family Dendrobatidae are masters of aposematic, or warning, coloration. Pictured are the funereal poison frog Phyllobates lugubris *(below)*, the strawberry poison frog Dendrobates pumilio *(bottom left)* and the orange and black poison frog D. leucomelas *(bottom right)*.

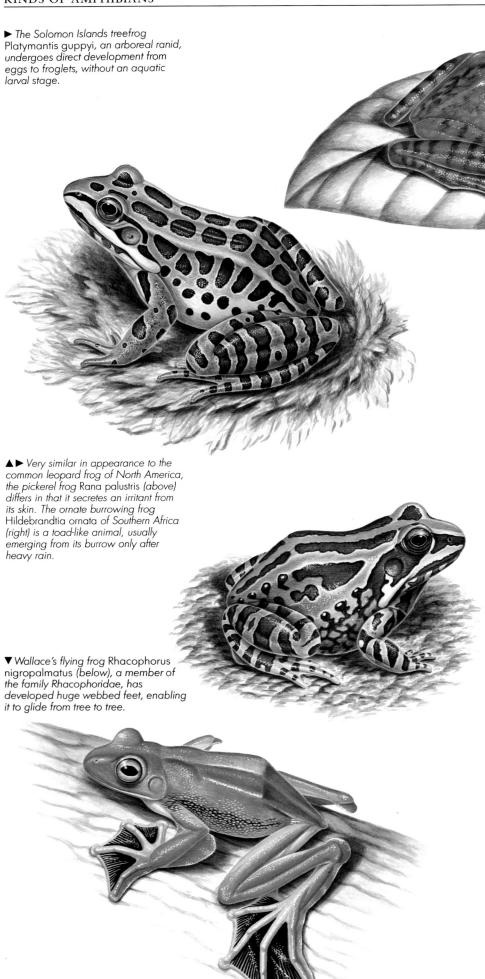

▶ The Solomon Islands treefrog Platymantis guppyi, an arboreal ranid, undergoes direct development from eggs to froglets, without an aquatic larval stage.

▲▶ Very similar in appearance to the common leopard frog of North America, the pickerel frog Rana palustris (above) differs in that it secretes an irritant from its skin. The ornate burrowing frog Hildebrandtia ornata of Southern Africa (right) is a toad-like animal, usually emerging from its burrow only after heavy rain.

▼ Wallace's flying frog Rhacophorus nigropalmatus (below), a member of the family Rhacophoridae, has developed huge webbed feet, enabling it to glide from tree to tree.

The northernmost species of frog is the common frog of Europe, *Rana temporaria,* nearly matched by the moor frog *R. arvalis.* These two, and the wood frog *R. sylvatica* of North America, range north of the Arctic Circle. The largest frog, the goliath frog *Conraua goliath* of West Africa, is a ranid, but the family runs almost the gamut of body sizes. Ranid frogs are most diverse in Africa, where there are 16 endemic genera, whereas Asia has about 12 genera. Europe and the Americas have only *Rana* species, with none held in common.

The genus *Rana* includes about one-third of the 650-odd species credited to the family. These are the classic "frogs" as compared to "toads": typically living in and on the margins of water, they are relatively smooth-skinned, powerful jumpers with long legs and extensive webbing on the feet. In most cases the eggs are laid in the water, followed by a tadpole stage lasting several weeks. The ranids present many examples of adaptive radiation. Frogs of the Asian genus *Amolops* have enlarged finger and toe disks (like those of treefrogs), which facilitate clinging to rocks beside the swift-flowing streams where their tadpoles live. The tadpoles, in turn, avoid being swept away by fixing themselves in place with a ventral sucker, rather than a sucker-shaped mouth as in other tadpoles of fast waters.

The burrowing-frog niche is exploited by the genus *Tomopterna,* called sand frogs in South Africa. Like frogs of other families with similar habits, these are squat, short-legged, somewhat wrinkled animals with a prominent digging tubercle or spade on each hind foot. Also in South Africa is the Hogsback frog *Anhydrophryne rattrayi,* named for the area where it was first found. This small frog lives in the forest leaf-litter, where the male digs a nest chamber in moist soil with his nose. The eggs develop directly to froglets, so there is no need for frogs of this species to frequent streams or ponds. At the aquatic extreme are small frogs of the genus *Occidozyga,* which live in swamps and pools in Asian rainforests. Their large, fully webbed hind feet are better adapted to swimming than to leaping.

Jane Burton/Bruce Coleman Limited

Because oceanic islands are typically barren of native frogs, the presence of two ranid species of the genus *Platymantis*—one a tree-dweller, the other terrestrial—in the Fiji islands is noteworthy. Frogs of this genus all live on islands, ranging from the southern Philippines to New Guinea, and eastward through the Solomon Islands before making the great jump to Fiji. Four closely related but morphologically diverse genera also inhabit the Solomons and share with *Platymantis* the direct mode of embryonic development. The history of how these frogs attained their present distribution can never be known for sure, but it undoubtedly involves passive distribution on islands moving very slowly over millions of years of tectonic activity. Numerous fossils from North America and Europe, none older than Oligocene (37 million years), are referred to the genus *Rana*.

Rhacophorid treefrogs
The rhacophorids, most of them treefrogs, are relatives of the largely aquatic and terrestrial ranids, and inhabit temperate and tropical parts of Africa and Asia, including Madagascar and Japan. This is a family of modest extent, with about 300 species in 12 genera, ranging in size from 15 to 120 millimeters (½ to 4¾ inches) body length. The flying frog *Rhacophorus nigromaculatus* of Southeast Asia is a member of this family.

The African genus *Chiromantis* has some interesting adaptations. In addition to the digital disks common to all arboreal frogs, *Chiromantis* has the inner two fingers opposable to the outer two, providing a firm grip on twigs. The frogs have unusual resistance to desiccation and can spend dry periods fully exposed. They lay their eggs in a tree above water. As the eggs are

▲ *Like many frogs of the temperate zone, the European brown or common frog Rana temporaria congregates at ponds in early spring to lay large communal masses of eggs. It has been suggested that one advantage of this behavior is that losses during cold weather are reduced: the black embryos readily absorb heat while the gelatinous envelopes provide insulation.*

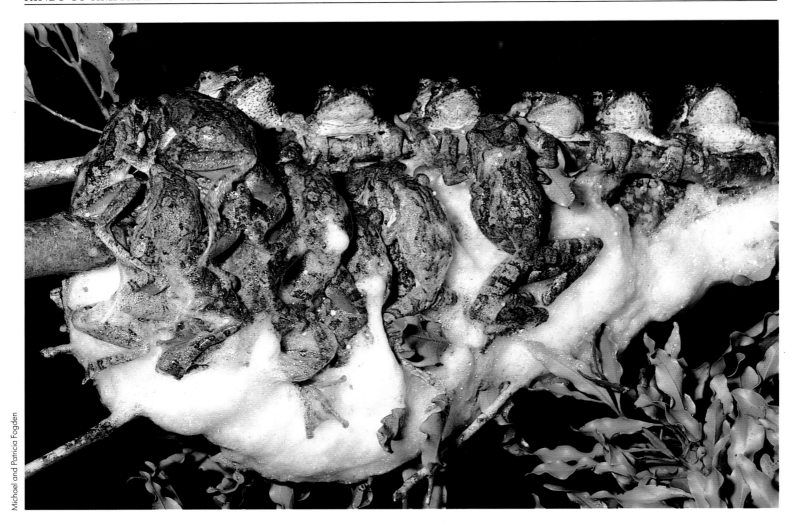

▲ *The gray treefrog* Chiromantis xerampelina *of arid southern Africa congregates in trees to mate in groups of up to 30 or so individuals, beating the eggs and seminal fluids with their feet to form a foam nest.*

▼ *Two members of the family* Hyperoliidae: *the painted reed frog* Hyperolius marmoratus *(left) and the Senegal running frog* Kassina senegalensis *(right). As its name suggests, the latter species walks or runs rather than hops.*

produced, the mating frogs use their feet to beat the eggs and accompanying liquid into a froth which hardens, protecting the developing eggs. Several pairs of frogs may work together building a communal nest. The larvae remain for a time in the nest before dropping into the water to complete development. Foam nests feature in the life history of most rhacophorid frogs.

Small tree-dwelling animals are unlikely candidates for fossilization, and there are no paleontological records for the Rhacophoridae—as indeed there are none for many other families.

Reed and lily frogs

The hyperoliids (Hyperoliidae) are another family of modest size—about 300 species, in 19 genera—related to the ranids. They are mostly small species, about 15 to 80 millimeters (½ to 3 inches) body length, living in Africa and Madagascar, with one endemic species on the Seychelles islands in the Indian Ocean. A modest adaptive radiation has produced tree-dwelling frogs and terrestrial frogs as well as a majority that climb but tend to remain for the best part of the time in low vegetation near water.

The genus *Hyperolius* includes half the species in the family. These small frogs live mostly in marshy or swampy areas, resting on reeds and sedges, on which many of the species deposit their egg masses. They are often boldly colored and patterned, with considerable individual variation in markings. This, and a general similarity of body form and structure, make it difficult to distinguish between the species, but a knowledge of the mating calls helps biologists to differentiate between them.

The genus *Leptopelis* includes both relatively large tree-dwelling frogs and burrowing frogs that rarely climb. Species of the genus *Afrixalus* lay their eggs on a leaf and then fold the edges of the leaf together, cementing them over the egg mass with secretions from the oviduct. Curiously, this may be done either in or out of water; in the latter case, the hatchling tadpoles must fall into the water to survive.

Hyperoliids are unknown as fossils.

Squeakers

The small family Arthroleptidae (about 70 species in seven genera), distributed in Africa south of the Sahara, is placed within the Ranidae by some authors. The voice of *Arthroleptis* species is the reason why people have named them "squeakers". These are small frogs that live on and within leaf-litter of the forest floor. The eggs, deposited in cavities or burrows in moist earth, undergo direct development; in some species the froglets are completely metamorphosed when they hatch, whereas in others the tail remains to be absorbed.

One arthroleptid is unique among all frogs. This is the so-called "hairy frog" *Trichobatrachus robustus* of Cameroon and Equatorial Guinea. In the breeding season, males develop vascularized hair-like structures on the flanks and thighs. They are reported to sit under water on egg masses in streams, and apparently the "hairs" serve to augment respiration through the skin, increasing the time the frogs can remain submerged.

No fossil arthroleptids are known.

Shovel-nosed frogs

The shovel-nosed frogs, family Hemisotidae, of Africa south of the Sahara, are 11 moderate-sized species (up to 80 millimeters, or 3 inches body length) in the single genus *Hemisus*. They are odd-looking frogs, round-bodied with short legs and a small pointed head with the tip of the snout hardened. Shovel-nosed frogs are burrowers, generally living in open country near

▼ *Common in the coastal lowlands of southern Africa, the waterlily frog* Hyperolius pusillus *lays its eggs in the cavities between overlapping waterlily leaves.*

Michael Fogden/Oxford Scientific Films

equatorial and southern Africa, and eastern India and Sri Lanka, through Southeast Asia to New Guinea and northern Australia. More than 320 species are recognized in 66 genera, the largest number of genera of any family of frogs. Microhylids comprise almost half the species of frogs in New Guinea and a sizable proportion in Madagascar, but are less significant elsewhere. Most are small frogs—several species are less than 15 millimeters (½ inch) body length—but others reach 80 to 90 millimeters (3 to 3½ inches). Morphology varies greatly, from rotund burrowers to typical treefrogs. A majority of the species live in moist tropical regions, but the evolutionary radiation of the group has placed species in arid habitats as well, and in a variety of terrestrial and arboreal niches. Some species are streamside frogs, but there do not seem to be any primarily aquatic microhylids.

Ground-dwelling microhylids live in both arid and humid tropical habitats. Among the most peculiar in arid regions are the rain frogs of the African genus *Breviceps*. (In arid regions they are likely to be seen only when it rains.) These frogs have small heads, short limbs, and round bodies, a shape accentuated by their habit of puffing up with air when disturbed. The arms are so short that the male cannot clasp the female around the body when mating (the usual method of maintaining contact between mating frogs), so instead the male and female become stuck together by secretions from.skin glands on the male's ventral surface, giving an effect not unlike that of two golf balls glued together. The eggs are laid in an underground chamber prepared or

▲ *The microhylids include a small genus of frogs, confined to Africa, with the unusual habit of laying their eggs in underground chambers. The embryo remains in the egg until metamorphosis is complete. These frogs inhabit arid regions and appear above ground only after rain. This is the common rain frog* Breviceps mossambicus.

pools, and are seldom seen above ground. The eggs are laid in an underground cavity, and the female remains with them until they hatch. Using her snout, she digs a burrow leading to water nearby, and the larvae then swim out and assume a more normal tadpole existence. Their bizarre appearance notwithstanding, these frogs are thought to be related to the ranids and are treated as a subfamily of Ranidae by some authors.

Microhylid frogs

The family Microhylidae occurs in the Americas from the southern United States to Argentina,

▼ ▶ *The Asian painted frog* Kaloula pulchra *(below) is a large microhylid that is now almost exclusively found associated with human settlements. The red-banded crevice creeper* Phrynomerus bifasciatus *(right, above) of southern Africa is a rock-dweller whose skin is rubbery in texture, giving this species its other common name of rubber frog. The eastern narrow-mouthed toad* Gastrophryne carolinensis *(right, below) is a much smaller species from the southeastern United States.*

enlarged by the female. The tadpoles do not feed but live on yolk provided in the egg, and the female remains with the nest until the tadpoles metamorphose and leave.

The moist leaf-litter of tropical forests is prime habitat for many microhylid species. Some burrow in the deep litter or soil and rarely emerge on the surface. Others come to the surface at night to wait for or actively seek food. The litter also serves as a daytime retreat for small climbing species that ascend into low vegetation at night to feed and advertise for mates. Microhylids with wide bodies but narrow, pointed heads commonly feed on termites and ants. Other species with more normal frog proportions have the catholic tastes of other frogs; one New Guinean species even eats other frogs. Many microhylids, notably in Madagascar and New Guinea, are tree-dwellers.

All the 120 or so species of microhylids found in New Guinea and Australia have direct development, skipping the tadpole stage. The tree-dwellers therefore do not need to leave the trees to seek pools or streams in which to breed, but may find appropriate arboreal sites—such as an epiphytic plant called the ant plant, which has a chamber that holds moisture.

Microhylid tadpoles differ from those of other families in features of their anatomy. With rare exceptions, they lack the horny beak and denticles ("teeth") of other frog larvae. In some species the mouth is formed into a funnel shape and is used, from below, in ingesting food from the water surface. (This adaptation occurs also in tadpoles of other families.)

The only pre-Pleistocene fossils of this family are from the early Miocene of Florida, about 24 million years ago. They are classified in the living genus *Gastrophryne* which inhabits this region near the present-day limits of the distribution of microhylids.

Four South American families

Some small families of frogs (lacking fossil records) are recognized not so much because of their distinctiveness but because their species cannot be fitted unambiguously into any of the larger groups. Four South American families—Allophrynidae, Brachycephalidae, Rhinodermatidae, and Pseudidae—are examples. *Allophryne ruthveni,* the only member of its family, is a small arboreal frog of northern South America variously considered most closely related to bufonids, leptodactylids, or hylids.

The three species of brachycephalid frogs live in the Atlantic forests of southeastern Brazil. They are very small frogs—*Psyllophryne didactyla* grows to less than 10 millimeters (less than ½ inch) body length, so is not only one of the two smallest frogs, but is also one of the smallest four-legged animals. The two species of the genus *Brachycephalus* are only slightly larger. Tiny frogs tend to have fewer

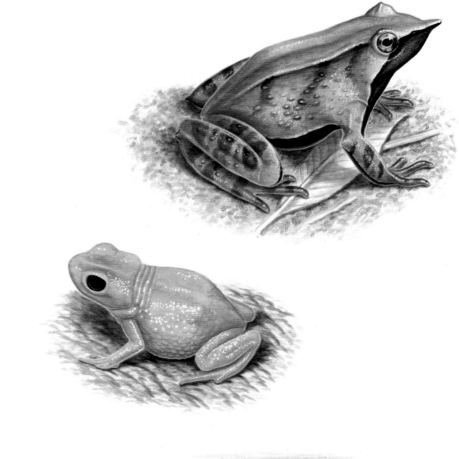

digits than usual, and brachycephalids, for example, have only three functional toes on each foot. At least one (and probably all three) species of brachycephalids have direct embryonic development, hatching as tiny frogs. In the family Rhinodermatidae the two species of the genus *Rhinoderma* are noted for their habit of oral brooding. They are small ground-dwelling frogs, about 30 millimeters (1¼ inches) long, found in the cool temperate forests of southern Chile and adjacent Argentina.

▲ *Three South American oddities: Darwin's frog Rhinoderma darwinii (top) is known for its unusual means of rearing young (see page 520); the gold frog Brachycephalus ephippium (center) is a diminutive burrower with a number of unique specializations for subterranean excavation; and the paradox frog Pseudis paradoxa (bottom), so named because the tadpoles may be up to four times the length of the adults.*

▶ *Restricted to the Seychelles in the Indian Ocean, the sooglossids are a small family whose relationship to other frogs is uncertain. The Seychelles frog* Sooglossus sechellensis *lays its eggs on land. After hatching, the tadpoles are carried on the back of the female until they metamorphose.*

R. A. Nussbaum

▼ *The Cape ghost frog* Heleophryne purcelli *belongs to a little known family confined to southern Africa. It is adapted for life in and around mountain streams, where the adhesive pads on its digits facilitate climbing on slippery rocks in fast-flowing water.*

The paradox frog *Pseudis paradoxa* gained its name because the tadpoles can reach remarkably large size, up to 250 millimeters (10 inches) in length, yet after metamorphosis the largest the frogs get is about 70 millimeters (2¾ inches). The five species in the family Pseudidae—two *Pseudis* and three *Lysapsus*—are almost totally aquatic, although an ability to survive dry periods buried

in mud has been reported for the paradox frog. The family ranges through tropical lowlands over much of northern and eastern South America from Colombia to Argentina.

Ghost frogs and Seychelles frogs

There are another couple of small families whose evolutionary affinities have been disputed. The so-called "ghost frogs", family Heleophrynidae, comprise the genus *Heleophryne* with five species confined to the Cape and Transvaal regions of extreme southern Africa. The common name may have been coined because one of the species is found in a place called Skeleton Gorge; certainly the frogs are not vaporous or otherwise ghostly.

Heleophrynid frogs are up to about 60 millimeters (2⅓ inches) long and are rather flattened, with prominently enlarged tips to the fingers and toes. They are therefore well adapted to fit into crevices and cling to rock surfaces along the cool, shaded mountain streams that are their habitat and where their tadpoles live. Like other tadpoles adapted for life in swift-flowing water, the tadpoles of ghost frogs have their mouthparts modified into a large sucking disk which allows them to cling to slippery rocks while feeding.

The suggestion that the heleophrynid frogs should possibly be classified within the Australian family Myobatrachidae implies a relationship going back many millions of years to when Africa and Australia were part of the Gondwanan supercontinent.

Evidence from chromosomes and behavior suggests that the family Sooglossidae, of the Seychelles, may also be related to the Myobatrachidae of Australia. There are three species—two in the genus *Sooglossus* and one in *Nesomantis*. They are small terrestrial frogs, up to 40 millimeters (1½ inches) body length, and they deviate from typical frog behavior in their method of breeding. Eggs are laid on the ground rather than in water and follow two modes of development: direct to small frogs in one species of *Sooglossus*; in the other species, tadpoles are carried on the back until they metamorphose. In

S. sechellensis, the tadpoles are carried not by the male, as is usual, but by the female.

CONSERVATION

A few species of frogs are listed as endangered by one agency or another, but it is not individual species so much as endangered habitats that need to be conserved. Destruction of rainforests in tropical regions has undoubtedly eliminated many species of frogs before they even became known to scientists. Wetlands in temperate areas and isolated sources of water in arid regions also merit special attention. Island faunas, too, are especially vulnerable to habitat destruction. Even where no specific cause can be identified, there are many instances of species apparently having disappeared or virtually so—for example, gastric brooding frogs in Australia and frogs of the family Ranidae native to southern California.

RICHARD G. ZWEIFEL

DISASTROUS INTRODUCTIONS

People have an unfortunate propensity for moving animals from their native area to exotic locations. Most attempted introductions probably fail, but success may create ecological disaster, or at least a lot of disturbance—consider the gypsy moth in North America and the rabbit in Australia.

Relatively few frog species have become established in places foreign to them, and most such introductions are probably benign. For example, formerly frogless Hawaii now has poison frogs from Central America among other species; some Australian frogs are established in New Zealand, which has only three native species; and a clawed frog of Africa *Xenopus laevis*, is spreading in southern California. But two widely introduced American frogs stand out from the rest: the bullfrog *Rana catesbeiana* and the marine or cane toad *Bufo marinus*.

The bullfrog is a large semi-aquatic species native to eastern North America. It is adaptable, prolific, voracious, and tasty. The last quality has resulted in its widespread introduction into western North America, where it competes with native frog species and may be a factor in their local extermination. It also occurs now in Puerto Rico, Italy (introduced in the 1930s), and Japan, and probably other regions as well.

The marine toad's native range is from extreme southern Texas to northern South America. Like the bullfrog, it is large, prolific, and adaptable, but being poisonous (and having relatively small hind legs), it is not eaten by humans. The excuse for its widespread introduction is control of insects that are agricultural pests. It occurs now on many Caribbean islands, Taiwan, the Philippines, New Guinea, and numerous islands in the Pacific, and

Jean-Paul Ferrero/AUSCAPE International

is spreading over northeastern Australia where it was introduced in 1935.

The marine toad's role in pest control is questioned, as a toad eats pest insects and beneficial insects indiscriminately. It has no natural enemies, and breeds all year round. Other aspects of the introduction are also clearly negative: dogs, cats, native mammals, birds, reptiles, and other amphibians can die from attempting to eat toads; and native frogs not only are preyed upon, but may be displaced ecologically.

▲ One of the factors in the cane toad's success is that it will eat almost anything that moves. It can handle surprisingly large prey, and has a voracious appetite.

PART THREE
KINDS OF REPTILES

TURTLES & TORTOISES

T he order Testudinata is divided into two suborders—Cryptodira (hidden-necked turtles) and Pleurodira (side-necked turtles)—and comprises about 250 species of turtles, tortoises, and terrapins distributed worldwide in tropical and temperate zones. They are the only reptiles that have a shell built into the skeleton, allowing them to more or less conceal themselves entirely within the shell. Of all the reptiles alive today, turtles and tortoises are not only the oldest forms but they have also changed very little in their 200-million-year history.

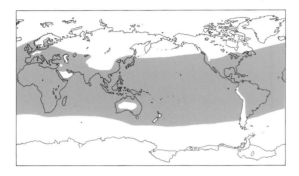

TWO MAIN LINEAGES

The oldest group of fossil turtles is known from the Triassic, about 230 million years ago. The turtles soon evolved into two main lineages which have survived until today: the cryptodirans or hidden-necked turtles (suborder Cryptodira) and the pleurodirans or side-necked turtles (suborder Pleurodira).

Cryptodirans can, by a vertical cobra-like bending of the vertebral column of the neck, draw the head directly into the shell (although in some modern species, only partial withdrawal is possible). In contrast, pleurodirans merely fold the head under the front edge of the upper shell by a sideways movement, either to the right or to the left. As well as these external characteristics, there are important differences in the structure of the skull and the skeleton.

Cryptodirans include sea turtles, which live in tropical and temperate oceans around the world, and also the majority of species that live on land or in rivers and lakes; they are found on all continents, although only one species reaches northern Australia. Pleurodirans are found only in Australasia, South America, and central and southern Africa.

In ecological terms, turtles have adapted to a wide range of habitats. They have successfully established themselves in dry landscapes—deserts, savanna, and plains—as well as grasslands, woodland, and mountains. Diverse species of freshwater turtles occur in still waters, such as ponds, and running waters, such as tropical rivers. A few species of freshwater turtles, like the true sea turtles, leave their aquatic habitats only to lay eggs, whereas other species are amphibious and regularly move about on land. There are a few species of land tortoises living in

► A green turtle *Chelonia mydas* at sea, Galapagos Archipelago. Marine turtles live at sea all their lives but come ashore on beaches to lay their eggs. Most species are endangered, having been heavily exploited by humans for food.

such dry landscapes that they are unlikely to encounter open bodies of water at any time throughout their lives.

CHARACTERISTICS IN COMMON

A turtle's shell consists of two parts: an upper part, called the carapace, and a lower part, called the plastron. Each part typically has an inner bony layer and an outer layer of horny plates. The visible layer is made up of large horny plates, but these actually cover a thicker layer of bony segments which makes up the true protective shell. Where the carapace meets the plastron, there are openings for the head, legs, and tail. The number and arrangement of horny plates tends to differ from species to species, and these arrangements are often useful in identifying species, although there are also individual differences within species.

Only three families lack horny plates on the shell: the softshell turtles, the Papuan soft-shelled turtle, and the leathery turtle. In these there is a thick leathery covering instead of the horny plates; in the softshell turtles this covering is flexible, at least at the edges.

A universal feature in modern turtles is the absence of teeth on the jaws. The oldest fossils

from the Triassic period did have very small teeth but these were on the palate, and the jaws themselves were toothless. Replacing the teeth in modern turtles and tortoises are horny ridges which cover the upper and lower jaws. In meat-eating turtles these horny ridges are knife-sharp and work like shears. In plant-eating species the outer edge of each horny jaw ridge is serrated, making it easier to bite off sections of hard woody plants, and there is often a serrated outer edge to the jaw in order to grip more readily.

All turtles possess strong limbs. Even the heaviest land turtle can lift its body off the ground when walking. In land turtles the fingers and toes have more or less grown together to form solid "clump-feet", whereas in aquatic freshwater turtles the individual digits are distinct and clearly recognizable, and many species have webbing between fingers and toes. The sea turtles are exceptional, in that their digits have again become fused during evolution to form paddle-shaped limbs with which they propel themselves through water, and in contrast to other turtles, the forelimbs are more strongly developed than the hind limbs. Sea turtles are also the only group that have to drag their body across the ground when they come ashore on sandy coasts to lay eggs.

▲ *The giant tortoise* Chelonoidis elephantopus *of the Galapagos Islands is among the largest of the land tortoises, second only to the Aldabra tortoise in size. Some individuals may exceed 1 meter (3¼ feet) in length, and approach 200 kilograms (about 450 pounds) in weight. But Galapagos tortoises remain vulnerable despite their size: rats prey on their eggs and hatchlings, and feral goats compete for scarce fodder.*

▶ *Great African tortoises Geochelone sulcata mating. Sniffing and butting are two common preludes to copulation among most tortoises: the sense of smell seems important in establishing age, sex, and readiness to mate, and a male repeatedly butts the female's shell until she responds by becoming quiescent, enabling him to mount.*

▼ *Marine turtles, like this loggerhead Caretta caretta, come ashore by night to lay their eggs, often in huge numbers. The female digs a hole in the sand above the high tide-line, deposits her clutch of 100 eggs or more, buries them, then returns to the sea. In due course, the eggs hatch and the hatchlings go to sea without further assistance from the female. Here the photographer has scooped away the sand at the rear of the egg chamber to expose the eggs.*

Ashod Papazian/NHPA

Jean-Paul Ferrero/AUSCAPE International

REPRODUCTION

All turtles lay eggs in a nest chamber, and the young develop in the eggs at a temperature corresponding with that of the surrounding sand or soil, without any further parental interest after the eggs are covered over. From the day they hatch, the young turtles must fend for themselves, their lives largely determined by the inherited instinctive behavior pattern of their species.

The incubation period varies to some extent with the microclimate within the egg chamber, but it is also genetically programed. The shortest incubation time seems to be about a month (for a softshell turtle); the average time is two to three months. The longest incubation time has been recorded in some land tortoises, which require one and a half years to incubate, although such times are reliable only when based on direct observation of the eggs or, less reliably, on breeding in captivity. In many species the hatchlings spend the winter in the egg chamber and appear for the first time the following spring, giving a false impression of long incubation. With the exception of a few larger species of sea turtles and freshwater turtles, the reproductive rate in turtles and tortoises is low. Clutches of the smaller species generally comprise only a few eggs (one to six), although many of the larger species may lay up to 150 eggs or more.

SIDE-NECKED TURTLES
Snake-necked turtles

Snake-necked turtles (family Chelidae) of South America, and Australia and New Guinea are well adapted to life in fresh water. The long neck of most species allows them to draw breath at the surface without exposing the rest of the body to potential predators, and they can stay underwater for lengthy periods while searching for food, particularly insects and their larvae, crayfish, tadpoles, and also small frogs and fish. Some species also feed on water plants, as well as fruits that fall into the water from surrounding trees.

The eight species in the genus *Chelodina*, known as Australian snake-necked turtles, live in Australia and New Guinea. Largest is the giant snake-necked turtle *C. expansa* of southeastern Australia, with a shell length of up to 42 centimeters (16½ inches); the total length of the head and neck is a further 31 centimeters (12¼ inches).

In seasonally dry areas some species will burrow deep into the mud at the bottom of lagoons and swamps to aestivate (remain dormant) until the next rain. When they need to cross dry areas between lagoons they generally wander during the night. They are helped by their ability to hold water in the anal sac, a structure that is also found in many land tortoises inhabiting desert regions.

The five species of short-necked turtles (genus *Emydura*) and three species of Australian snapping turtles (genus *Elseya*) have shorter necks, more or less normal in length. They live in Australia and New Guinea, mostly in flowing water, and can swim extremely well. Like many other stream-dwellers, the short-necked turtles are often seen sunbathing in the early morning, lying beside or on top of one another in favorable positions on the river bank, and if disturbed dive rapidly into the water.

As recently as 1980 a new genus and species was discovered in eastern Australia, the Fitzroy turtle *Rheodytes leukops*, found only in the Fitzroy River drainage in Queensland. It was given the name "*leukops*" because of the remarkably white iris of the eye.

In southwestern Australia, near the city of Perth, lives the rarest and most threatened turtle species in the world, the western swamp tortoise *Pseudemydura umbrina*. Probably fewer than 40 of

these little creatures survive in the wild (in a small reserve), but a captive breeding program in Perth Zoo has recently had significant success. If the original habitat can be made secure from predators and human interference, the wild population can be supplemented from the captive breeding program.

South American members of the family Chelidae include both snake-necked and short-necked groups, and the two species of American snake-necked turtles (genus *Hydromedusa*) are superficially very similar to the Australian *Chelodina*. Among the stream-dwellers there are at least eight species in the toad-headed genus *Phrynops*, the best known being Geoffroy's side-necked turtle *P. geoffroyanus* of Brazil and Paraguay. The largest is the spotted-bellied side-necked turtle *P. hillarii*, with a shell length of up to 44 centimeters (17¼ inches) and weighing up to 1.2 kilograms (2½ pounds); it lives in the Rio Paraná and Rio Paraguay and their tributaries in eastern South America. New *Phrynops* species have been found in recent years, and it seems likely that others are still to be discovered.

In the streams of the Chaco region in south-central South America, which can be dry for months at a time, live smaller turtles in the genus *Acanthochelys* whose ecology is similar to that of the pond turtles (family Emydidae), described later. One species, the Chaco side-necked turtle *A. pallidipectoris*, is notable for the long horny spurs on its upper thighs. It measures only

▼ The eastern snake-necked turtle Chelodina longicollis *is common in swamps and wetlands of southeastern Australia. It is a member of the family Chelidae, a group of aquatic and semi-aquatic turtles with representatives in Australasia and South America. The head is retracted into the shell by horizontal (rather than vertical) folds of the neck.*

Jean-Paul Ferrero/AUSCAPE International

head and neck, with flaps of skin on the sides, look remarkably like a fallen leaf. Thus camouflaged, the matamata lies motionless in shallow pools in the forest or in slowly flowing streams, occasionally lifting its snout to the surface to breathe. Fish do not recognize it as a predator and swim carelessly close to its mouth.

Helmeted side-necked turtles
Side-necked turtles of the family Pelomedusidae occur only in South America, Africa, and Madagascar. Despite the small number of species in South America, they are so abundant that they are regarded as characteristic animals of South

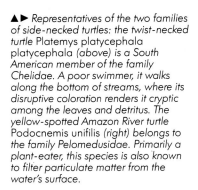

▲▶ *Representatives of the two families of side-necked turtles: the twist-necked turtle* Platemys platycephala platycephala *(above) is a South American member of the family Chelidae. A poor swimmer, it walks along the bottom of streams, where its disruptive coloration renders it cryptic among the leaves and detritus. The yellow-spotted Amazon River turtle* Podocnemis unifilis *(right) belongs to the family Pelomedusidae. Primarily a plant-eater, this species is also known to filter particulate matter from the water's surface.*

18 centimeters (7 inches) in shell length, and when the streams dry up it buries itself deep in the mud until the rainy season begins. The remaining two or three species in this genus live in eastern South America.

In northern South America, in the river systems of the Amazon and the Orinoco, are two members of the family Chelidae that have a number of unusual features. The twist-necked turtle *Platemys platycephala*, measuring only 18 centimeters (7 inches) in shell length, lives in shallow, slow-flowing streams in the rainforest. Its bright yellow/orange/brown coloring is reminiscent of turtles in the forests of Southeast Asia and probably serves a similar purpose: as a disruptive coloration that camouflages the turtle on the leaf-strewn forest floor or stream bed.

An equally effective camouflage is achieved by the matamata *Chelus fimbriatus* but using a different strategy. Its shell, which is up to 45 centimeters (17½ inches) long, is flattened and ridged, so that it looks like a piece of bark, and its

American wildlife. The giant South American river turtle *Podocnemis expansa* and the yellow-spotted Amazon River turtle *P. unifilis* are not only conspicuous because of their size—the shell length exceeding 100 centimeters (3¼ feet) and almost 70 centimeters (2¼ feet) respectively—but also because of the enormous number that seek out nesting spots on the sand banks of the Amazon and Orinoco rivers.

Unlike many other large freshwater species, *Podocnemis* turtles have a predominantly vegetarian diet.

The Madagascar big-headed side-necked turtle *Erymnochelys madagascariensis* differs very little from its South American relatives, and its presence on this island off the southeast coast of Africa is dramatic evidence of Madagascar's origin as a part of a large southern continent, Gondwana, which included South America, Africa, Australia, and Antarctica. The family is represented in Africa by widespread species of the genus *Pelusios,* all of which have a hinged front section of the plastron. Ranging in size from 12 to 45 centimeters (5 to 17½ inches) long, these roundish turtles have strong-smelling musk glands whose secretions deter potential predators. They live in both flowing and still waters, preferring to hide in the mud, where they find their food—mollusks, worms, and insects. As the water dries up, they bury themselves in the mud and aestivate there until the rainy season.

The African helmeted turtle *Pelomedusa subrufa* is similarly widely distributed over eastern and southern Africa. It also occurs in the outermost southwest tip of Asia in the Yemen, on the Arabian Peninsula, and so is the only side-necked turtle occurring in Asia. In habits and form it differs little from the *Pelusios* turtles.

HIDDEN-NECKED TURTLES
Alligator turtles (snapping turtles)

The family Chelydridae today contains only two species living in North and Central America. They are remarkable for their long tail, which has large shield-like scales similar in appearance to those of crocodiles. They also have a remarkably large head which cannot be completely withdrawn into the shell. If an alligator turtle is turned on its back it can be seen that the plastron (the shell covering the belly) is reduced in size and cross-shaped. The limbs cannot be withdrawn into the shell but

can only be drawn tightly up against it.

The American snapping turtle *Chelydra serpentina* is found from southern Canada to southern Ecuador and is the smaller of the two species, with a shell length of up to 47 centimeters (18½ inches). It consumes a predominantly animal diet, including salamanders, fish, frogs, and smaller birds and mammals. Much of the time is spent in the water, although these turtles also like to sunbathe in the mornings on the banks of streams and swamps.

The alligator snapping turtle *Macroclemys temmincki* of the southeastern United States, from the Mississippi Valley southwards to Texas and Florida, is the largest freshwater turtle in North America. The record size is a shell length of 66 centimeters (26 inches), and a weight of 114 kilograms (250 pounds); the head of this particular individual was 24 centimeters (9½ inches) long and about as broad.

As well as a large powerful head, the species has a hooked beak and a circle of large keeled scales around each eye, which makes the eyes appear larger. There is also a series of long soft scales on the neck, giving it a prickly appearance. The shields of the carapace are strongly overlapping like roof tiles, while the flanks have an additional row of horny shields, considered to be a primitive characteristic.

◄ *The helmeted turtle* Pelomedusa subrufa *is widespread across eastern and southern Africa, occurring also in Madagascar and even the extreme southwestern corner of the Arabian Peninsula. It favors temporary floodwaters as well as ponds and streams, and frequently moves from one area to another as its shallow ponds dry up, or alternatively aestivates in the mud until the rains return.*

Anthony Bannister/NHPA

▶ *The alligator snapping turtle Macroclemys temmincki of the eastern United States takes its name from the pronounced bony ridges on the carapace as well as from its swift and savage bite. It has a pink, fleshy appendage on its tongue, which it moves in its wide-open mouth to lure fish and other unwary prey to within reach of its snapping jaws.*

J.A.L. Cooke/Oxford Scientific Films

▼ *Female olive ridleys Lepidochelys olivacea massing to lay their eggs on Nancite Beach, Costa Rica.*

David Hughes/Bruce Coleman Limited

The alligator snapping turtle is more aquatic than its smaller relative, the American snapping turtle. Large older animals rarely leave the water, and then only the females for egg-laying. They eat virtually everything they can capture—even large snails and mussels are unable to resist their strong jaws—and are especially well adapted to catching fish using a mechanism found in no other turtles. The tongue has a reddish-colored worm-like appendage which, when the mouth is held wide open underwater, moves about in a life-like manner. Any fish that tries to eat the decoy worm will itself become a victim of lightning-fast jaws.

Hardback sea turtles

The seven species of sea turtles (family Cheloniidae) are survivors of a much larger group which reached its greatest diversity during the Jurassic and Cretaceous periods, 200 to 65 million years ago.

All recent sea turtles of the family Cheloniidae are fairly uniform in structure, form, and life history. However, one additional species is so distinctive as to warrant its recognition in a separate family—the leathery turtle is the sole survivor of an otherwise extinct family and is discussed separately.

The Cheloniidae all have rather flat shells with a complete covering of large horny plates. The forelimbs are more strongly developed than the hind limbs, a feature distinguishing them from freshwater turtles. The bony shell is less substantial than that of freshwater turtles; between the bony plates of both the carapace and the plastron are broad gaps, filled in by fibrous skin, and these in turn are covered by the strong horny plates. In this way the very heavy shell of land turtles, which is unnecessary for a turtle spending most of its life at sea, is much reduced without loss of structural support and stability.

Sea turtles leave the water only to lay their eggs, so after leaving the beach as hatchlings, the males spend their entire life in the sea. They often sunbathe at the surface, drifting or resting on floating fields of seaweed, or in shallow water left by the receding tide on coral reefs. There is some evidence that females return to lay their eggs on the same beaches as those from which they themselves hatched; certainly females return year after year to the same beaches to nest. Unfortunately, their homing instinct has allowed humans to predict their arrival for breeding, with the consequence that sea turtles have been almost exterminated from many of their breeding grounds and their existence is threatened globally.

Together with the flatback turtle *Natator depressus*, the green turtles (genus *Chelonia*)—with one species, *C. mydas*, in the Atlantic, Indian, and western Pacific oceans, and *C. agassizii* in the eastern Pacific—are mostly sought for their flesh,

but to some extent green turtles are also hunted for the horny plates of their shells.

Ridley sea turtles are also hunted for their meat. The olive ridley *Lepidochelys olivacea* is found throughout much of the Atlantic and Pacific, while Kemp's ridley *L. kempii* is found only in the Gulf of Mexico and warm waters of the Atlantic. The latter species seems to use only a single nesting beach on the eastern coast of Mexico, making it especially vulnerable to overexploitation and extinction. Whereas green turtles attain a shell length of up to 150 centimeters (5 feet), ridley sea turtles rarely have a shell length of more than 70 centimeters (2$\frac{1}{3}$ feet).

Somewhat larger (up to 90 centimeters or 3 feet shell length) is the hawksbill turtle *Eretmochelys imbricata*, which seems to be restricted to warm tropical seas. The horny shields of the upper shell are beautifully marbled or flamed, and are much sought after for ornaments, putting the species at great risk from overexploitation.

The largest member of the family is the loggerhead turtle *Caretta caretta*. It reaches a shell length of up to 213 centimeters (7 feet), although the average is about 150 centimeters (5 feet). Its

▼ *Green turtles* Chelonia mydas *mating. These marine turtles migrate sometimes thousands of miles to breed. The sexes rendezvous to mate at sea near the nesting beaches, but no pair bond is formed, and both males and females may mate with others several times during the mating period.*

Alby Ziebell/AUSCAPE International

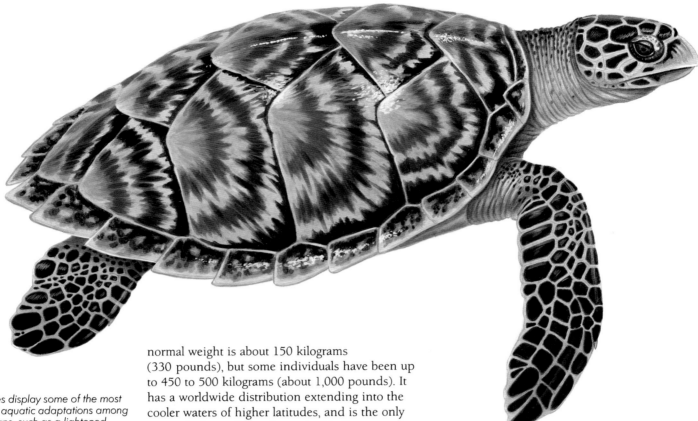

▲ Sea turtles display some of the most specialized aquatic adaptations among the chelonians, such as a lightened, hydrodynamically-shaped shell and large paddle-like forelimbs with reduced claws. The attractively marked Pacific hawksbill Eretmochelys imbricata bissa has been the source of commercial "tortoise shell", used in the manufacture of items from eyeglass frames to haircombs, with devastating consequences for wild populations.

normal weight is about 150 kilograms (330 pounds), but some individuals have been up to 450 to 500 kilograms (about 1,000 pounds). It has a worldwide distribution extending into the cooler waters of higher latitudes, and is the only sea turtle that still breeds on the Mediterranean coast. Fortunately its flesh is not eaten nor is its shell commercially useful. A major threat is tourism, as people and boats disturb the nesting beaches and discourage breeding, while large numbers are drowned in fishing nets. Consequently, it is not surprising that despite its large clutch size (females usually lay more than 100 eggs in a clutch and may lay several clutches in a season) loggerheads are as endangered as other sea turtles.

Leathery turtle

The largest turtle alive today is the leathery turtle *Dermochelys coriacea*, also known as the luth or leatherback, the only living representative of the family Dermochelyidae. One exceptional individual has been reported with a shell length of almost 2.5 meters (8¼ feet) and weighing about 860 kilograms (1,900 pounds), but specimens

SAFETY IN NUMBERS

The most hazardous period in the life of a sea turtle is when it leaves the nest and crosses the open beach on its way to the sea, running the gauntlet of predatory birds and crabs. But there is safety in numbers—the more small turtles there are, the better the chance they will not be picked off one by one. When the turtles hatch, they dig simultaneously toward the surface, gradually moving the chamber upward as the sand is deposited beneath them. Finally, and usually at night, they burst forth and scuttle for the waves, where they are safe from their land-based enemies, but not from ocean-dwelling predators.

▶ Flat-back turtle hatchlings Natator depressus.

Jean-Paul Ferrero/AUSCAPE International

with shell lengths of more than about 1.5 meters (5 feet) are uncommon. The powerful forelimbs project as paddles with a greater span than the shell length itself, so the sight of a fast-moving leathery turtle in the open sea is very impressive. The large head is characterized by big eyes and a conspicuous hooked beak. Unlike the other sea turtles, whose shells are covered by horny plates, the shell of this species is covered only by a leathery skin. There are seven tubercular longitudinal ridges on the carapace (upper shell).

The leathery turtle feeds largely on jellyfish, but its diet may also include mollusks, echinoderms, and crustaceans. Apparently fish are rarely eaten. It is a cosmopolitan species, occurring in tropical and temperate seas throughout the world and extending into the colder waters of higher latitudes. However, like other sea turtles the females return to the same nesting beaches over and over again to lay their eggs, usually the same beaches at which they themselves hatched. Leathery turtles are often washed up exhausted and die on the cool coasts of Europe or North America, so it seems certain that the species can either maintain its body temperature above that of the surroundings for longer periods or that it can tolerate cooler conditions than can other sea turtles.

Softshell turtles

The softshell turtles of the family Trionychidae include species that display a wide range of adaptations to an aquatic existence in rivers and lakes, but they are surprisingly uniform in form and habits.

Softshell turtles are characterized by not having the usual horny shields of the epidermis, but instead a leathery skin. While the central part of the carapace has a bony layer, the outer ring of

solid bones has been lost during evolution. A subsequent strengthening of the plastron (lower shell) has evolved in many species through the development of dermal bones which are not attached to the bony shell and are visible as coarse, hard spots in the plastron. The leathery skin, with its smooth surface, flexible edge, and especially the very elastic hind third of the shell, is well adapted to fast and energy-efficient swimming in open water, as well as movement in the muddy bottom of streams and lakes.

Softshell turtles "settle" themselves on the bottom with an undulatory movement, where they lie hidden from predators, although they are generally able to defend themselves effectively with knife-sharp horny jaws, usually concealed under swollen lips. They have a strongly vascularized throat, which is able to extract oxygen from the water and which can allow them to delay having to come to the surface to breathe. The leathery skin can also exchange oxygen with the surrounding water, so softshell turtles tend to be able to stay underwater for longer periods than other turtles. Most softshell turtles are strictly carnivorous and feed on mollusks, crustaceans, aquatic insects, worms, frogs, and fish. Even smooth flatfish are successfully grasped and held fast before they are consumed. A few species also eat fruit and aquatic plants. Despite their many adaptations to aquatic life, softshell turtles in northern latitudes regularly come onto land to sunbathe.

Softshell turtles are now classified in about 15 genera and are found in North America, Africa, and Asia but not in South America or Australia, although they occur as fossils in Australia. The North American genus *Apalone* has three species, one of which, the spiny softshell turtle *A. spinifera,*

Jane Burton/Bruce Coleman Limited

▲ *The softshell turtles of the family Trionychidae have a characteristic habit of lying partly buried in the muddy bed of rivers and ponds, using their ability to obtain oxygen through their skins to reduce the need to come to the surface to breathe.*

▼ *Like the sea turtles, softshell turtles have independently evolved numerous adaptations for an aquatic existence, including a reduction in shell armor, large webbed feet with few claws, and a snorkel-shaped snout for breathing while remaining beneath the surface. The eastern spiny softshell Apalone spinifera spinifera is a colorful species of southeastern North America.*

▲ *The southern loggerhead musk turtle* Sternotherus minor minor *is a small freshwater turtle of southern North America. As individuals of this species mature and graduate from a juvenile diet of insects, the head and jaws grow disproportionately large to accommodate the adult's main diet of mollusks, hence its common name. The "musk" refers to a smelly fluid expelled by these turtles when disturbed.*

has a distribution extending from southern Canada to northern Mexico.

The largest number of softshell turtles are found in Asia. In Asia Minor is the Euphrates softshell turtle *Rafetus euphraticus*, whose range overlaps with the African softshell turtle *Trionyx triunguis*, a large species (up to 95 centimeters, or 37 inches shell length) from southern Turkey and along the Mediterranean coast as far as Egypt and Somalia, as well as in west Africa from Mauritania to northern Namibia. In some areas it inhabits the brackish estuaries of larger rivers.

The Asiatic softshell turtle *Amyda cartilaginea* occurs throughout much of the East Indies, including Java, Sumatra, and Borneo. With a shell length of 70 centimeters (27½ inches) it is quite large, but it is exceeded by two other species in this region: the narrow-headed softshell turtle *Chitra indica*, which reaches 115 centimeters (45 inches) in shell length, and the Asian giant softshell turtle *Pelochelys bibroni*, with a shell length of 129 centimeters (51 inches).

In Africa and India live several softshell species whose plastron has large flaps of skin that conceal the turtle's feet. Two African species (in genera *Cyclanorbis* and *Cycloderma*) inhabit the rivers and some lakes in tropical and central Africa. The genus *Lissemys* is found in India but also has two species in Ceylon and Indo-China. The Indian flapshell turtle *L. punctata*, like many other turtles, can apparently survive the dry season by burying itself deep in a riverbed or the bank of a drying stream or swamp.

Papuan softshell turtle

The Papuan softshell turtle or pitted-shell turtle (family Carettochelyidae) is externally similar and most closely related to other softshell turtles, in that instead of having its shell covered with horny plates, there is a layer of thick leathery skin, but unlike softshells in the family Trionychidae, the shell of the Papuan softshell turtle is not flat and plate-like but is domed with a ridged keel along

its midline—a shape found in many other swamp turtles. The bony skeleton in the shell is completely supported and has a strong bony margin which provides a solid support and a bridge with the plastron. While these are smaller than in most "hardshell" turtles, all of the normal bony structures are present.

The family Carettochelyidae is represented today by a single species, *Carettochelys insculpta*, which is found only in New Guinea and tropical northern Australia. It has a long trunk-like snout with tubular nostrils which allow the turtle to breathe at the surface without putting the rest of its head out of the water. It is largely vegetarian, preferring the fruits of bog plants and trees that grow beside its aquatic habitats, such as pandanus and figs. It will, however, take animal food such as mollusks, crustaceans, and worms. Like the sea turtles, it uses its forelimbs rather than its feet to propel itself through the water.

Mud and musk turtles

The family Kinosternidae includes small, cryptic turtles confined to North and South America. The average size is 15 to 20 centimeters (6 to 8 inches) shell length. All species have a solid carapace, which is often characterized by three long keels and is covered by strong, occasionally overlapping, horny shields. The plastron in most species is large, and two distinct hinges allow the front- and hind-most portions to move in such a way that the turtle can completely close its shell front and back. In some species only the front section of the plastron is hinged, allowing only partial closure of the shell, while in a few others the plastron is reduced to an immovable cross-shaped structure which offers little protection to the limbs and other soft parts of the body.

Probably the best known genus is that of the mud turtles (genus *Kinosternon*), with about 15 species occurring from the United States, through Central America to northern South America. Most of these are inconspicuous, brown-colored turtles, which spend the greater part of their day on the bottom of streams and lakes feeding on mollusks. In the mornings they leave the water to bask in the sun and achieve their preferred body temperature. Many species, especially the smaller ones, climb shrubs and even trees beside the water.

A second genus, the musk turtles (*Sternotherus*), has only four species, confined to the central and southern United States. All have a reduced plastron. They are often referred to as "stinkpot" turtles because of the extraordinarily strong musky smell they exude when captured. This odor is not confined to *Sternotherus* but occurs in all members of the family and many other aquatic turtles. The habits of *Sternotherus* are similar to those of other kinosternids. Associated with their small size they produce fewer eggs: one to five is the normal clutch size.

In Central America there are members of two further genera which, because of differences in anatomy and cell structure are sometimes classified in their own family, the Staurotypidae. Firstly, there are the cross-breasted musk turtles (genus *Staurotypus*): the large Mexican giant musk turtle *S. triporcatus*, almost 40 centimeters (16 inches) in shell length, and the Chiapas giant musk turtle *S. salvinii*, to 25 centimeters (10 inches) shell length. As the common name implies, their plastron is reduced to a strong, bony cross, but with a large head and a sharp horny beak they can nevertheless defend themselves effectively. Both species range from Mexico to Honduras or El Salvador.

In Mexico and in Guatemala there is another member of this group, the narrow-bridged musk turtle *Claudius angustatus*, whose shell length is only 15 centimeters (6 inches). It has a conspicuously large head with an impressive beak, and is unable to conceal it within the shell. When it bites, this turtle holds on ferociously and can cause a painful wound, while at the same time releasing copious amounts of a smelly secretion from the cloaca.

Central American river turtle

The sole living representative of family Dermatemydidae is the Central American river turtle *Dermatemys mawi* of Mexico and northern Central America. It looks like a typical large freshwater turtle, although its stronger, flatter shell has a conspicuous row of additional shields on the bridge, where the carapace meets the plastron. This impressive turtle, whose shell length is 65 centimeters (25½ inches), inhabits both fresh and brackish water in rivers, lagoons, and estuaries. It tends to seek out the warmer upper levels of deeper water, letting itself drift while raising its body temperature. When disturbed it dives quickly into deeper water.

Its food consists mainly of aquatic plants, and the horny edge of each jaw is serrated to assist it in cutting hard woody plants. However, it will also take any available animal food. They leave the water only to lay eggs, up to 20 in a clutch, twice a year, and usually close to water.

New World pond turtles

In their conquest of freshwater habitats, apparently two similar groups evolved in parallel. The pond turtles (many of which are called terrapins) of the New World (family Emydidae) and the pond turtles of the Old World (family Bataguridae) apparently shared a common ancestor, and most researchers today believe that that common ancestor was also shared by land tortoises (family Testudinidae).

The Emydidae, with eight genera, are most diverse in southern North America. A few species

extend north to Canada, others occur in the Caribbean and through Central and South America. A single species extends into South America, and one species, the European pond turtle *Emys orbicularis*, is the only representative of this family in Europe, North Africa, and the Middle East.

Pond turtles are distinguished by having a full bony shell covered with horny plates and, in some genera, well-developed hinges on the plastron which can completely close the shell. They also have well-developed limbs with webbed feet. Most are semi-aquatic and occur in swamps, rivers, and even coastal lagoons. Some are more terrestrial and live in woodlands far from water. In fact, no other turtle family lives in such a wide variety of habitats.

The best known types are the three genera of ornamented turtles—the painted turtles (*Chrysemys*), the sliders (*Trachemys*), and the cooters (*Pseudemys*)—so-called because of their brightly colored shells, heads, and limbs. One of the most beautiful is the painted turtle *Chrysemys picta*, which occurs as a number of subspecies from southern Canada to the far south of the United States. It has a bright red or yellow design on the brown carapace and the head. The shell length is about 25 centimeters (10 inches).

Scarcely less spectacular are the different geographic forms of the slider *Trachemys scripta*, especially those that have large ocellate patterning on the sides of the carapace—giving rise to another common name, the peacock-eyed turtle. The species is found from the northern United States, throughout Central America, to South America. Other *Trachemys* species are found on islands in the West Indies, and members of the related genus *Pseudemys* are distributed throughout the eastern United States. Male ornamented turtles are generally smaller than the females. They also have noticeably longer claws

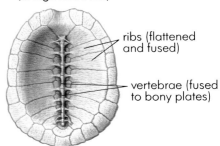

ribs (flattened and fused)

vertebrae (fused to bony plates)

VIEW FROM BELOW

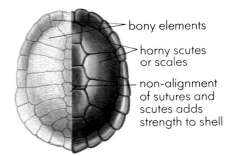

bony elements

horny scutes or scales

non-alignment of sutures and scutes adds strength to shell

VIEW FROM ABOVE

▲ The shell of chelonians has two main parts: the upper shell, or carapace, and the lower shell, or plastron. The shell is constructed of interconnected bony plates which, in the carapace (shown above), include the expanded and fused ribs, with the backbone fixed permanently in place. Both the carapace and plastron have a second covering layer of large horny plates, or scutes. The suture lines between scutes do not align with those of the bony plates, adding to the shell's strength.

▼ A male painted turtle Chrysemys picta belli (family Emydidae) of North America.

▶ *The eastern box turtle* Terrapene carolina carolina *is a terrestrial member of the family Emydidae, and has evolved a tortoise-like appearance. The plastron in this species is hinged, allowing full closure of the shell with head, limbs, and tail withdrawn, giving the turtle a box-like appearance. This individual is a male, as indicated by its red-colored eyes.*

▼ *The sawback or false map turtle* Graptemys pseudogeographica *of North America. The sexes differ so greatly in size in this and related genera that some separation in habitat and diet is often evident, with the strongly vegetarian females tending to live in deeper waters, while the much smaller males favor a more carnivorous diet in the shallows.*

on the forelimbs which they use in a complex dance ritual for the females prior to mating; they approach the female from the front and rhythmically stroke both sides of her head with these long claws, stimulating her to mate.

Closely related to the ornamented turtles and occurring in the same areas of the eastern and southern United States are ten species of map turtles (genus *Graptemys*). In size they range from 15 to 30 centimeters (6 to 12 inches) shell length, and males are again much smaller than females. A characteristic of this genus is the presence of humps or ridges on the central shields of the carapace. Map turtles are not as colorful as ornamented turtles, although their heads have decorative patterns, and in many species the outer shields of the carapace are beautifully decorated with eye-shaped markings. All these turtles like to bask in the sun, especially in the early morning when they congregate on the banks of ponds, ditches, and rivers to raise their body temperature, although always prepared to plunge into deeper water when alarmed. Adults tend toward a vegetarian diet, whereas the young tend to be more carnivorous, but there is considerable variation in diet between species and even between the sexes and individuals.

Allied to the ornamented turtles is the chicken turtle *Deirochelys reticularia*, with a conspicuously long neck. Like the diamondback terrapin *Malaclemmys terrapin*, it is highly sought after for its flesh. The diamondback terrapin prefers brackish water in bogs, lagoons, and estuaries along the eastern and southern coast of the United States, where there is a broad range of food including crustaceans, mollusks, and aquatic plants. It has been exterminated in many areas as a result of being collected for food.

The box turtles (genus *Terrapene*), of eastern and southern United States and Mexico, have a plastral hinge which allows the shell to close completely. Most box turtles live on land, the Carolina box turtle *T. carolina* in moist deciduous forests and grasslands, whereas the ornate box turtle *T. ornata* prefers the drier, sandy landscapes of the prairies. Only the Coahuilan box turtle *T. coahuila* is strictly aquatic; it occurs only in the Cuatro Ciénegas basin in northern Mexico. Blanding's turtle *Emydoidea blandingi*, with its plastral hinge, resembles the box turtles but with its long neck also resembles *Deirochelys* species. Like true pond turtles it prefers larger bodies of standing water in the central regions of the United States, and it was for many years erroneously included in the genus *Emys*.

The European pond turtle *Emys orbicularis* is geographically and evolutionarily distinct. This medium-sized pond turtle (up to 28 centimeters, or 11 inches, shell length) is semi-aquatic and is found in North Africa and in most of Europe, eastwards to the Aral Sea. Although it eats both plant and animal material, the vast majority of its diet is animal.

The last genus of New World pond turtles, *Clemmys*, contains four species. The best known and one of the smallest freshwater species is the spotted turtle *C. guttata*, which lives in small pools and ditches. The largest of the four is the wood turtle *C. insculpta*, which has a shell length of 23 centimeters (9 inches) and lives in deciduous forests of eastern North America, where it is largely independent of water.

Land tortoises

Strictly speaking, the term "land tortoises" applies only to members of the family Testudinidae. Land tortoises are found in Europe, Africa, Asia, and all of the Americas. The best known are the European species in the genus *Testudo*, especially the Greek or Hermann's tortoise *T. hermanni* and the spur-thighed tortoise *T. graeca*, which occur almost continuously throughout the countries bordering the Mediterranean Sea. Two additional species have more limited distributions: the marginated turtle *T. marginata* is found only in Greece and Sardinia, while the Egyptian tortoise *T. kleinmanni*, the smallest species (up to

15 centimeters, or 6 inches, shell length) occurs from Libya to Israel. Depending on the latitude at which they occur, these tortoises have a short or long hibernation period during which they bury themselves in the ground. They depend heavily on seasonal supplies of fresh herby plants, for all land tortoises are predominantly vegetarian, although they will eat insects, worms, and mollusks, and even carrion and the dung of hoofed animals. Greek and spur-thighed land

Stan Osolinski/Oxford Scientific Films

▲ *Ornate box turtles* Terrapene ornata *mating. Copulation presents special difficulties for land turtles with high, strongly domed carapaces. Several turtles have evolved spurs and other aids for enabling the male to stay in position; the box turtles of North America have movable shells, often used by the female as a clamp to hold her mate's hind feet in position as he falls on his back to mate.*

Hellio & Van Ingen/NHPA

◀ *A European pond turtle* Emys orbicularis *feeds on an introduced sunfish. An omnivorous and semi-aquatic species, this turtle feeds mainly on frogs, worms, mollusks, fishes, and even rodents and small birds. In the northern parts of its range it buries itself in mud to hibernate for much of the winter.*

tortoises have been valued as pets for many years, but this trade has threatened the survival of numerous southern European and North African populations.

The richest place on Earth for land tortoises is Africa. The genus *Geochelone* predominates, with the African spurred tortoise *G. sulcata* of the Saharan region reaching 75 centimeters (30 inches) shell length and 80 kilograms (175 pounds) in weight. Like the gopher tortoises of America this species survives dry periods by burrowing deeply into the soil. In hot, dry regions all land tortoises are active only in the morning and late afternoon. During the heat of the day they spend their time resting in the shade of shrubs and trees or in burrows in the earth. If exposed to the heat of the midday sun the bulky body, covered with horny plates and a massive bony shell, would quickly overheat and the tortoise perish.

The leopard tortoise *G. pardalis* is widely distributed in the southeast of the continent and is regarded as one of the typical animals of the savanna. This species has the longest incubation time for its eggs: up to 460 days. The eggs of all

land tortoises have a calcified shell which is resistant to damage and rapid dehydration. To enable them to dig a nesting chamber in the often-hard ground, female land tortoises may release urine and water (stored in their anal sac) to saturate the ground and make digging easier.

Also widely distributed in the dry regions of Africa is Bell's hinge-back tortoise *Kinixys belliana*. This and two other species in the genus have a transverse hinge on the back of the carapace which allows the rear part of the shell opening to be completely closed as protection from many predators—although it is not safe from the bite of the hyena, which is able to completely penetrate the shell. The two other species live in rainforests of Central and West Africa, feeding mostly on worms, snails, and insects on the moist forest floor.

Southern Africa has at least two genera of land tortoises with restricted·ranges. The South African land tortoises, *Psammobates*, with shell lengths of 14 to 24 centimeters (5½ to 9½ inches), depend on specific food plants. The smallest land tortoises are five *Homopus* species (9 to 15 centimeters or 3½ to 6 inches, shell length) found in dry regions of southernmost Africa. One of these, *H. bergeri*, was thought to be extinct until recently rediscovered in a very small area.

Another species recently rediscovered after almost 90 years is the most remarkable land tortoise of East Africa, the African pancake tortoise *Malacochersus tornieri*. Hatchlings have a closed and arched shell, but the adult tortoise is not only very flat but also very soft. Between the bony plates in the shell there are open spaces, which increase in proportion as the tortoise grows larger (to 17 centimeters, or 6½ inches, shell length). The horny plates covering the shell develop normally and reveal little of the reduced bony layer beneath. This tortoise lives in the mountains of East Africa and is an able climber. With its flexible shell and by inflating its body with air, it is able to wedge itself in rock crevices· to avoid being pulled out by predators. Females

▼ *One of the most beautifully marked tortoises, the radiated tortoise Asterochelys radiata (below) belongs to a genus with only two species, both confined to Madagascar, and both endangered. The appropriately named pancake tortoise Malacochersus tornieri (below, right) of east Africa is an unusual species, well adapted to its rocky habitat. The shell is not only remarkably flat, but is flexible as well, allowing the tortoise to squeeze into narrow crevices, and wedge itself in by inflating its lungs and expanding its shell.*

lay one or two eggs several times each year, and with this small reproductive rate the species is especially vulnerable to collection for the pet trade.

There are four land tortoises endemic to Madagascar. Best known is the radiated tortoise *Asterochelys radiata* found in dry regions of the island's southwest. Unfortunately the local people regard it as a delicacy. In the mid-west of Madagascar lives the Angonoka or northern Madagascar spur tortoise *A. yniphora*, whose males have a long spur on the plastron which is used to drive off rivals during the mating season, rather as antlers are used in many mammals. Similar spurs are found in males of the South African bowsprit tortoise *Chersina angulata* and the American gopher tortoises (genus *Gopherus*).

Two small species are found in the south of Madagascar: the Malagasy spider tortoise *Pyxis arachnoides*, so-named because of the spider-web design on its yellow carapace, and the Madagascar flat-shelled spider tortoise *P. planicauda*, which grow to about 12 centimeters (4¾ inches) shell length. For most of the year they are inactive, living in burrows underground, and when the rainy season begins they have only a few weeks to feed, mate, and lay eggs.

On the Seychelles, the Mascarene Islands (Mauritius and Réunion), and Aldabra live the largest land tortoises on Earth. The Aldabra tortoise *Aldabrachelys elephantina*, with a shell length of up to 130 centimeters (51 inches), is larger than its distant relative the Indefatigable Island tortoise *Chelonoidis elephantopus nigrita* from the Galapagos. Although these two giant species belong to different evolutionary lineages, their shells are structured in the same way: the bones of the shell have a honeycomb structure, which encloses many small air chambers. If the bony shell were solid it would be difficult for these giant tortoises to carry around such weight.

The nearest relatives of the Galapagos giant tortoise live on the South American mainland. The South American red-footed tortoise *Chelonoidis carbonaria* and the South American yellow-footed tortoise *C. denticulata* inhabit forests of tropical regions. With shell lengths of up to 80 centimeters (31½ inches) they are not significantly smaller than the smaller races of Galapagos Island tortoises, but because the shell is slender and elongated they never achieve the weight of Galapagos tortoises of similar shell length. A third species, the Chaco tortoise *C. chilensis*, lives in the grasslands of Argentina and Paraguay. Despite its scientific name, it does not occur in Chile.

North America is home to the gopher tortoises (genus *Gopherus*), from Florida to California and southwards to northern Mexico. The fifth species, *G. lepidocephalus*, was only recently discovered in the south of Baja California but may already be nearing extinction. The other four species, which are also endangered, fortunately receive statutory

Tui de Roy/Oxford Scientific Films

protection. Found from desert and semi-arid regions to moister woodlands, all gopher tortoises are burrowers which tend to spend the heat of the day below ground. They are especially active in the early morning and evening, when they feed on various plants, like all land tortoises finding food largely through a combination of smell and sight.

The Asian brown tortoise *Manouria emys* is something of a giant among the land tortoises of Southeast Asia, reaching 60 centimeters (23½ inches) shell length. This flat, plain brown tortoise occurs in rainforests, especially with open bodies of water, in which it likes to bathe and where it feeds on aquatic plants and animals. While other land tortoises bury their eggs in the ground, this species scratches together a mound of soil and leaves in which it buries its eggs. The nest may hold up to 50 eggs, and because of its elevation above the forest floor it is protected from flooding. A more colorful relative is *M. impressa*, of northern Indo-China. It has a much flatter shell, in brown or black patterned with bright yellow or orange, effectively camouflaging the tortoise on the leaf-strewn floor of the tropical forest. Three species of *Indotestudo* and various species of *Geochelone* are found in India. The Indian star tortoise *G. elegans* is one of the most beautiful of all land tortoises, with its dark striped patterning on each shield of the carapace. It lives in savanna and dune habitats on the Indian mainland and Sri Lanka.

▲ *Darwin's finches (genus* Geospiza) *picking ticks from a Galapagos giant tortoise. Despite the massive appearance of the enormous carapace, the shells of these large tortoises are internally honeycombed—presumably to reduce weight—and surprisingly fragile and prone to injury.*

▲ *Two colorful Southeast Asians (family Bataguridae): the spined or cogwheel turtle* Heosemys spinosa *(top) has remarkable spines, most prominent in young individuals. In addition to the physical deterrent they provide, the spines, combined with the turtle's coloration, help to conceal the animal in leaf-litter. The Malayan snail-eating turtle* Malayemys subtrijuga *(above) is an aquatic species of slow or still waters that feeds primarily on mollusks.*

Old World pond turtles

The Old World pond turtles (family Bataguridae) form the largest family of turtles, with approximately 21 genera. Apart from a single genus (*Rhinoclemmys*), in Central and South America, all are found in Europe, North Africa, or Asia. Not surprisingly there are many parallels with the family Emydidae (described earlier) in form, structure, and biology. Indeed species of the two families can be so similar that only finer anatomical differences can be used to distinguish between them, but fortunately, the families have separate geographic distributions (apart from the exceptions mentioned here).

River turtles have a distinctive solid shell and strong, fully webbed feet. As a general rule they are

the largest of the pond turtles, the Malaysian giant turtle *Orlitia borneensis*, for example, having a shell length of 80 centimeters (31½ inches). These turtles are characteristic animals of the larger rivers of India, Indo-China, and Indo-Malaysia, including Borneo. Several of them, including *Callagur*, *Hieremys* and some *Batagur* species, are also found in brackish waters. They tend to be omnivorous when young but strictly vegetarian when adult.

The painted terrapin *Callagur borneoensis*, of southern Indo-China, Sumatra, and Borneo, often lays its eggs on the same coastal beaches where sea turtles make their nests. On hatching, the young first enter the sea and live there briefly until they make their way into estuaries and rivers. The painted terrapin is also remarkable for the distinctive color pattern adopted by males during the mating season: a red crown on the normally gray-olive head, almost silvery-white on the nape, and a distinctive brightening of colors on the carapace. This kind of seasonal color change is common in lizards but rare in turtles.

Apart from large river turtles the family Bataguridae also includes a number of small species, which are also mostly aquatic, leaving the water only to sunbathe or lay their eggs. Among these are the eyed turtles (genus *Morenia*) with two species in Bangladesh and Burma, three species of the Indian roofed turtles (genus *Kachuga*), the Chinese stripe-necked turtle *Ocadia sinensis*, the spotted pond turtle *Geoclemys hamiltoni*, which ranges through the whole of northern India, from Pakistan to Bangladesh, and the black marsh turtle *Siebenrockiella crassicollis* from Indo-China and the Greater Sunda Islands. Most of these aquatic turtles are largely vegetarian in diet, although there are some notable exceptions—for example, the Malayan snail-eating turtle *Malayemys subtrijuga*.

In contrast, the four species in the genus *Mauremys* are decidedly amphibious. The Caspian turtle *M. caspica* and Mediterranean turtle *M. leprosa* have an omnivorous diet and live in streams in arid regions and mountainous areas around the Mediterranean Sea. The streams tend to dry up completely during the summer, so the turtles migrate overland, often long distances, to find new sources of water. In extreme cases they are forced to bury themselves to avoid desiccation and wait until the next rain. Similar habits are found in the Japanese turtle *M. japonica*, which lives in the southern highlands of Japan.

Another amphibious group are the hinged tortoises of India and southwards through the Indo-Malayan archipelago to the Greater Sunda Islands. Some specialists classify them in two distinct genera, *Cuora* and *Cistoclemmys*; other people place them in a single genus, *Cuora*. All have a plastral hinge which allows the turtle to withdraw its head and limbs into a completely enclosed shell. Some species are truly amphibious and spend much of their time in water. Others, like the Indochinese box turtle *Cuora galbinifrons*, in the mountain forests of southern China and Vietnam, are essentially terrestrial and can survive long periods without water. This species has a brilliantly colored shell whose warm yellow, orange, and brown pattern provides excellent camouflage on the leafy forest floor.

Other turtles in the family are even more terrestrial. In southern China and Indo-Malaysia, in the same habitat as *C. galbinifrons*, lives the keeled box turtle *Pyxidea mohouti* and the black-breasted leaf turtle *Geoemyda spengleri*—the latter species extending to Sumatra, Borneo, and the islands of Japan. Both of these species have a serrated edge to the keels that extend back along the carapace, so as well as being camouflaged by the light and dark brown color, the serrated edges of the shell give the turtle a leaf-like appearance when still.

These various forest dwellers, which ecologically are "land turtles", have many other representatives. Probably the best known is the bizarre spiny turtle *Heosemys spinosa*, which when young has a circular shell edged with regular spines. In adults the spines are less conspicuous, except on the hind edges of the shell.

The only genus in this family that is found in the Americas is *Rhinoclemmys*, with seven to nine species in Central and South America. Some are terrestrial, others amphibious. All of them are more or less brilliantly colored, especially the head and sometimes the shell itself, which may have beautiful eye-shaped markings, as in the painted wood turtle *R. pulcherrima*, or spots, as in the Mexican spotted wood turtle *R. rubida*. All land turtles, both the Asiatic and the American, are omnivorous, although predominantly herbivores. They feed on fallen fruits, but generally avoid mushrooms and other fungi.

Big-headed turtle

The East Asian big-headed turtle *Platysternon megacephalum* may be closely related to the land tortoises (family Testudinidae) or the Old World pond turtles (Bataguridae), but sometimes is classified in a separate family, Platysternidae.

It lives in southern China and northern and central Indo-China. The big-headed turtle is rarely more than 20 centimeters (7¾ inches) in shell length. Its conspicuous, large and powerful head cannot be fully retracted under the shell, nor can the long tail with its large horny scales. The flattened shell, although of normal proportions and development, is too small to fully enclose the fleshy parts of the turtle. On either side of the bridge that links the carapace and plastron is an additional row of large shields, a primitive characteristic that also occurs in alligator turtles (snapping turtles) of North and Central America and the Central American river turtle.

The big-headed turtle lives in cool, fast-flowing mountain streams and, even though it is not a powerful swimmer, is well adapted to grasp and climb among the large boulders on the stream bottom. Specimens kept in captivity have drowned if the water in their tank is too deep, and others that have escaped from their aquaria have been found not on the floor but near the ceiling, having climbed up the curtains!

By day the big-headed turtle lies hidden among stones, but at night it emerges to hunt snails, crabs, and fish. With its beak-like jaws it can grasp its prey tightly, and with very strong jaw muscles it can even bite through the thick shells of its prey. Its reproductive rate is low, and only one to two eggs per clutch have been recorded. However, in its natural environment the big-headed turtle has few predators, offsetting its low reproductive rate. The greatest risk to this species is from collectors for the pet trade.

FRITZ JURGEN OBST

◄ *The big-headed turtle, here represented by the North Vietnamese race* Platysternon megacephalum shiui, *is a unique Asian turtle. Found in cool mountain streams it is primarily aquatic but a poor swimmer, preferring to walk along the bottom. It is a good climber, however, and is occasionally seen basking on the lower branches of streamside bushes or trees. Its huge head cannot be withdrawn into the shell and is consequently heavily armored, as is the long tail.*

LIZARDS

KEY FACTS

ORDER SQUAMATA
SUBORDER SAURIA
• 26 families • 420 genera
• 4,300 species

SMALLEST & LARGEST

Monito gecko *Sphaerodactylus parthenopion*
Head–body length: 17 mm (⅔ in)
Total length: 34 mm (1⅓ in)
Weight: 0.12 g (⁴⁄₁₀₀₀ oz)

Komodo monitor *Varanus komodoensis*
Head–body length: 75–150 cm (2½–5 ft)
Total length: 170–310 cm (5½–10¼ ft)
Weight:35–165 kg (77–364 lb)

CONSERVATION WATCH
!!! There are 15 critically endangered species listed: *Abronia montecristoi; Diploglossus anelpistus;* Montserrat galliwasp *Diploglossus montisserrati;* Smith's dwarf chamaeleon *Bradypodion taeniabronchum;* Paraguanan ground gecko *Lepidoblepharis montecanoensis;* Roosevelt's giant anole *Anolis roosevelti;* Fiji crested iguana *Brachylophus vitiensis; Cyclura carinata;* Jamaican iguana *Cyclura collei;* Anegada rock iguana *Cyclura pinguis;* Ricord's iguana *Cyclura ricordi;* Hierro giant lizard *Gallotia simonyi;* Bermuda rock skink *Eumeces longirostris;* Allan's lerista *Lerista allanae;* Saint Croix ground lizard *Ameiva polops.*
!! There are 15 species listed as endangered.
! 66 species are listed as vulnerable.

Lizards today occupy almost all landmasses except Antarctica and some Arctic regions of North America, Europe, and Asia. During the extinctions that occurred at the end of the Cretaceous period, 65 million years ago, lizards survived but dinosaurs and other large reptiles did not. Other surviving reptiles—turtles, crocodilians, and tuatara—have not evolved into as many different forms. Indeed, since the only other large group of living reptiles, the snakes, evolved from lizards, it may be said that more than 95 percent of living reptiles are the descendants of the early lizards. Probably fewer than 800 species of dinosaurs existed during the entire span of the group's existence (about 140 million years), compared with more than 4,300 lizard species existing today, regardless of the number of fossil species.

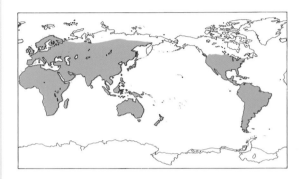

SMALL SIZE AND SUCCESS
Part of the reason for the much greater diversity of lizards is their small size; few living lizards exceed 30 centimeters (1 foot) in total length, and only a handful exceed 1 meter (3¼ feet). A given geographic region can support a greater diversity of smaller animals than it can of large animals. This is related both to the lower demand for food and other resources per individual and the diversity of microhabitats available to small animals. In a forest, for example, many habitats may be available to a typical lizard (the soil, leaf-litter, holes in tree trunks, treetops) and each might be occupied by a different lizard species. On the scale of a large dinosaur, however, the forest as a whole might represent just a single habitat. The partitioning of a habitat and its resources is well demonstrated in groups of lizards that have many species.

Also related (in part) to small size, most lizards have limited ability to spread geographically. Mountain ranges and expanses of water, such as rivers, lakes, and seas, are significant barriers for lizards and have promoted speciation (the evolution of new species), resulting in many forms that occur only in a small geographical area.

ORIGIN AND EARLY EVOLUTION
Lizards and their evolutionary offshoots—the snakes and amphisbaenians—constitute a group known as the Squamata, or "scale reptiles", and are representatives of diapsid reptiles. The lineage leading to lizards (the Lepidosauromorpha)

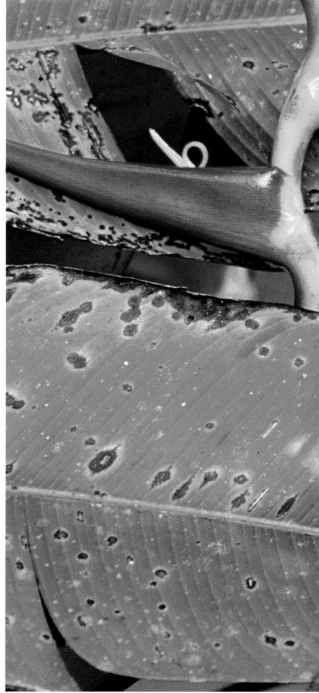

diverged from the other major lineage of diapsids (the Archosauromorpha including crocodilians, dinosaurs, and birds) during the Permian period, 285 to 245 million years ago. Lizards probably first appeared in the Triassic, 245 to 200 million years ago, but fossils definitely assignable to living lizard families are not known until the middle to late Cretaceous, about 120 to 100 million years ago. However, the presence of fossil lizards closely resembling living groups from the Jurassic, 200 to 145 million years ago, suggest that most of the body plans typical of the living lizard families were established almost 200 million years ago.

Lizards as a whole are closely related to the Sphenodontia, or tuatara, represented today by two species in New Zealand. Both groups share such features as the presence of a hooked fifth metatarsal bone, a part of the foot that acts as a heel in locomotion. Lizards, however, differ in skull structure, the skull often being highly mobile; when the quadrate bone (the point of attachment of the jaw) rotates forward, it pushes against the palate, which in turn lifts the lizard's muzzle. This modification of the feeding apparatus has also contributed to the success of lizards as a lineage. Lizards also differ from tuatara in that they have hemipenes, another feature they share with snakes (discussed later).

▼ *Dwarfed by the heliconia leaf on which it sits, a male green basilisk* Basiliscus plumifrons *basks in a Costa Rican rainforest. The huge variety of different lizard species is largely a function of their small size: the smaller an animal is, the greater the range of microhabitats potentially available to it.*

◀ Day-active reptiles inhabiting deserts need behavioral mechanisms to shed heat as well as absorb it. In the sand dunes of the Namib Desert in Africa, a sand-diving lizard Meroles anchietae lifts two feet alternately while balancing on the remaining two in order to reduce the transfer of heat from the hot sand—an activity that might well be termed "thermal dancing".

Most lizards have eyelids and external ear openings (snakes do not), but these features reflect the retention of primitive characteristics, rather than the evolution of conditions peculiar to lizards. Thus, lizards can be defined as those squamates that lack the derived and often highly specialized features that define snakes and amphisbaenians.

THERMOREGULATION

Lizards, like other reptiles, do not have the ability to regulate their body temperatures physiologically. The metabolic heat they produce is minimal and is quickly dissipated to the environment through the skin. Under normal conditions, the body temperature of an exposed lizard quickly approaches that of the surrounding environment.

At low temperatures lizards face a dilemma. It is necessary to expose the body to sunlight in order to warm up. Yet at such temperatures the abilities of lizards to move effectively are reduced, and they are vulnerable to predation by mammals and birds, which are not as constrained by external temperatures. Because they cannot run well at low temperatures, some lizards show behavioral compensation of a sort. At very low temperatures lizards are unable to respond to most stimuli, but at slightly higher though still not ideal temperatures, lizards are often highly aggressive, relying on less energetically expensive bluffs or bites than on escape. Lizards that can endure tail loss (autotomy) are usually more prone to do so at low temperatures in order to distract predators and gain the additional time needed to escape.

It is common for lizards to begin their morning basking by exposing the head only or by basking in the shelter of a crack or crevice. Heat uptake through solar radiation may be enhanced by the darkening of the skin. This is accomplished by the hormonally controlled dispersal of melanin (the same substance found in human skin) within the cells of the skin. Once the animal has heated to an acceptable level it may then emerge to search for food or engage in other activities. Nocturnal lizards may either bask (usually at protected sites) or obtain heat through thigmothermy, the transfer of heat by contact with a warm surface. Many nocturnal geckos spend the daylight hours under bark or thin flakes of stone, absorbing heat by conduction from their surroundings.

Lizards must also cope with the stress induced by high temperatures. Although the preferred body temperatures for many lizards are high (up to 42°C, or 107°F, for sustained periods), temperatures only slightly higher may be fatal. Most lizards limit their activities to certain periods of the day. Those living in open tropical environments or in deserts may be active only in the morning and afternoon so that they avoid the heat of midday. Others remain active throughout the day but shuttle back and forth between sun and shade, regulating their body temperature to

◀ (Opposite) Despite the tropical situation of the Galapagos Islands, the seas surrounding them are cold. Insofar as thermoregulation is concerned, marine iguanas, Amblyrhynchus cristatus, accordingly have the best of both worlds—living along the shoreline they can bask on rocks in the tropical sun, warming themselves quickly after foraging in the cool water.

Gunther Deichmann/AUSCAPE International

▲ A number of species of desert-inhabiting lizards evade the torrid extremes of heat at the surface by burrowing in the sand, like this burrowing skink Lerista labialis in Australia's Simpson Desert.

▼ Among lizards, two strategies have evolved to deal with sand: some species, like sandswimmers, burrow in loose sand and tend to have reduced limbs, while the inhabitants of the firmer, windward faces of sand dunes often excavate their burrows with large webbed feet, like this gecko Palmatogecko rangei of the Namib Desert munching its cricket prey.

picking up heat radiated from the surface. *Meroles anchietae*, a lacertid lizard which inhabits the sand sea of the Namib Desert in Africa, is especially noteworthy in this regard. It raises a foreleg and the opposite hind leg simultaneously, balancing on the other two legs for a few seconds to allow the skin of the feet to cool. The lizard then alternates the lifting and lowering of the diagonal pairs for as long as it is exposed to uncomfortably hot sand.

For many lizards, both extreme heat and extreme cold may be encountered at different seasons. In temperate regions, winter is generally a period of inactivity, and metabolic rates may drop to minimal sustainable levels as temperatures remain so low that basking and foraging activity cannot take place. In warmer zones such as temperate to subtropical deserts, lizard activity in winter may be restricted to the middle of the day, when the temperature is suitable. Typically animals in such areas remain inactive during midday in the summer, when a lethally high temperature is likely to be encountered.

The length of the basking period is controlled hormonally by the pineal gland, a brain structure that lies beneath the roof of the skull and is often known as the "third eye". In many lizards, especially those that are primarily day-active, the pineal gland is not covered by bone but lies beneath a foramen (or window) in the skull, covered by a translucent scale.

within a degree or two of their preferred value. Many desert lizards avoid the excessive heat by burrowing, because at a depth of only a few centimeters the temperature may be drastically cooler than at the surface; and at a depth of 30 centimeters (12 inches) it may be as much as 35°C (95°F) cooler. Other lizards, such as geckos, remain in burrows or other protected sites during the day and only venture out at night when desert temperatures may be more equitable. Lizards may also orient the broadest part of their bodies into the wind in order to lose heat by convection. Still others move their bodies off the ground to avoid

DESERT ADAPTATIONS

Although lizards are found in almost all habitats, they are often thought of as desert animals. In countries such as Namibia in southwest Africa, and in Australia, where arid habitats predominate, lizards are especially numerous.

Deserts impose a number of challenges to lizards; high daytime temperatures must be endured, and low humidity and a lack of free water tend to result in dehydration. Water may be obtained entirely through the food consumed or may come from one of several other sources. In the coastal deserts such as the Namib in Africa and the Atacama in South America, fog originating over the ocean provides the water used by lizards and other animals and plants. A number of lizards living in arid zones are able to effectively use rain as a source of drinking water. Normally, light rains quickly soak into the dry desert soil and become unavailable to surface-dwelling reptiles, but at least two types of agamid lizards have developed a mechanism for salvaging such a resource: the Australian thorny devil *Moloch horridus*, and the Asiatic toad-headed agamid *Phrynocephalus mystaceus* have their scales arranged in such a way that water is channeled to the mouth by capillary action. A similar mechanism to collect fog moisture or dew may also function in these or other lizards.

Anthony Bannister/NHPA

Dehydration is combated by features that slow down the rate of water loss from the body. All lizards have a covering of keratin (a substance similar to that of human fingernails) which acts as an effective barrier to water. Lizards are also able to reduce water loss through excretion, because they are able to limit the amount of filtrate (and thus water) passed out of the kidney. The nitrogenous wastes produced by lizards consist mainly of uric acid, which is relatively insoluble and can be concentrated and stored with little need for dilution by water.

In arid regions with shifting sands, special demands are placed on lizards. While many lizards inhabit deserts, most occupy specialized microhabitats such as rock islands or clumps of vegetation. The open, shifting sands of dunes are especially challenging and have been exploited by a limited number of iguanids, geckos, lacertids, skinks, and girdle-tailed lizards. Two general strategies for living and moving in the dunes are employed. Most dune-dwellers, such as the sandfish *Scincus scincus* of Asia and North Africa and the sand lizard *Uma notata* of North America, utilize loose sand, which does not contain enough moisture to allow the formation of tunnels or burrows. The sand lizard generally buries itself for protection from predators and high temperatures, whereas the sandfish actually moves freely through the substrate in an action referred to as "sand swimming". Both animals have countersunk lower jaws to prevent sand from entering the mouth, and both have valved nostrils and modifications of the ear openings, also to exclude sand. In the case of the sandfish, smooth scales reduce friction as it moves through the sand, while in the sand lizard, small granular scales may perform a similar function. Enlarged scales form fringes at the borders of the toes in these and other ecologically similar lizards; the fringes increase the surface area of the feet for progression through sand or for initial burial by shimmying or lateral undulations of the body.

A much smaller number of lizards are specialists inhabiting the more compact, windward dune faces. These lizards build open tunnels and so lack features such as valves over their nostrils needed for sand burial. Instead their modifications are for sand excavation. The gecko *Palmatogecko rangei* has very large, webbed feet for exactly this purpose.

REPRODUCTION AND LONGEVITY

All bisexual lizard species (those with both males and females) exhibit internal fertilization. Lizards have a horizontally aligned vent, which is the exit of the cloaca, a common vestibule for the digestive and uro-genital tracts. Male lizards possess a pair of intromittent organs, the hemipenes. When not in use the hemipenes lie adjacent to the cloaca within the base of the tail.

Ken Griffiths/NHPA

Austin James Steven/Bruce Coleman Limited

▲ Lizards exhibit the full spectrum of reproductive possibilities, from egg-laying to live birth, but egg-laying tends to be the rule. In some, eggs are laid in chambers excavated in the soil then covered over. Here an Australian eastern water dragon Physignathus lesueurii puts the finishing touches to her clutch.

◀ A hatching Seychelles green gecko Phelsuma abbotti patiently works to tear itself free of the prison of its eggshell. Although most lizards produce leathery-shelled eggs, the eggs of gekkonine geckos like this species are covered by a hard, brittle shell.

During sexual activity one hemipenis is everted by the action of muscles and fills with blood. The fully everted structure may be relatively simple or highly complex, as in many chameleons, which exhibit folds and spines that differ from species to species. In copulation, which follows courtship behavior, only a single hemipenis is inserted into the female's cloaca, and the sperm travel along a groove in the hemipenis. Retraction of the hemipenis is accomplished by drainage of the blood sinuses and activation of retractor muscles that invert the structure as it is withdrawn. In at least several species of anoles, males tend to alternate which hemipenis is used when mated repeatedly.

Oviparity (egg-laying) is the predominant mode of reproduction among lizards, but viviparity (live-bearing) has evolved on many different

▲ *Kicking and biting, two rival male frill-necked lizards* Chlamydosaurus kingii *battle it out. The enormous frills are used to bluff the opponent into believing the owner is bigger than he really is.*

occasions (at least 45) and is seen in nine families. The Brazilian skink *Mabuya heathi* even has a mammalian-like placenta; the fertilized egg is minute and through gestation grows 38,000-fold, entirely as a result of maternal nourishment. In fact, most lizards fall on a continuum between live birth and oviparity, as about half of the developmental period, on average, occurs within eggs retained in the female's body. In most instances, embryos of viviparous reptiles obtain their nutrition from yolk, and live birth is an extension of egg retention. The placental condition of *M. heathi* is one of only a few instances of extensive maternally provided nutrition in lizards.

Live-bearing has both advantages and disadvantages. On the one hand, viviparous lizards can protect their young throughout the prenatal development period. They are also able to regulate the temperature of the developing young at a level greater than that provided by a nest, decreasing the time required for development. The evolution of viviparity in lizards is probably related to the occupation of cold climates, which would strongly favor these advantages. On the other hand, the added weight and bulk of embryos must be carried by the female for an extended period of time. This may substantially reduce the speed and agility of the adult and make her more prone to predation. Increased basking and thus exposure in gravid females, as seen in some Australian skinks, may also increase the risk of predation. Likewise, the need to carry the young for the full term may

prevent the production of multiple clutches in a given season. In geckos of the genus *Naultinus*, for example, gestation may last as much as eight months, limiting an individual to only one litter (of two young) every one or two years. Lifetime reproductive output may thus be quite low for livè-bearers, but survival rates are higher.

Although lizards do not live as long as some other reptiles, some species have impressive longevity. A slow worm, *Anguis fragilis*, lived in captivity for more than 50 years, and some wild populations of geckos are known to include individuals over 20 years old. On the other hand, a few, mostly very small species may live only one or two years. Life-spans of more than ten years are probably exceptional for most species in the wild, although large varanids and helodermatids may commonly exceed this—the Komodo monitor *Varanus komodoensis* has a life-span of as much as 50 years. Most reptiles continue to grow long after reaching sexual maturity. Nonetheless, studies of very old individuals suggest that growth ceases at some point, and growth rates typically slow down long before this point is reached.

COMMUNICATION

Many lizards spend much of their life as solitary individuals, yet at times they communicate with one another through a set of highly stereotyped behaviors. These include aggressive behaviors directed by males at other rival males, and courting behaviors between the sexes. Other types of social interactions are also probably important in some species, but our knowledge of these is rudimentary. Males of virtually all iguanids and agamids engage in patterns of head-bobbing and/or push-ups as a means of communication. Comparable information is conveyed by movements of the tail in other lizards. The day-active semaphore geckos (genus *Pristurus*) have a laterally compressed tail that is used to signal members of its own species across significant distances in their relatively open habitat. Many reptiles are brightly colored, and such patterns are frequently important cues in mate recognition and selection. In some male lizards, particularly iguanians, color displays may be enhanced by the erection of crests or dewlaps. Members of the genus *Anolis* show species-specific color patterns of the dewlap that are exposed when the animal displays. Females may exhibit bright colors as well. In the keeled earless lizard *Holbrookia propinqua*, females exhibit hormonally controlled color changes associated with their reproductive cycle. Such changes may signal that they are receptive to males and are usually accompanied by changes in female behavior.

Autarchoglossan lizards rely on color as well. The orange head color of breeding males of the *Eumeces* skinks elicits aggressive behavior from males of the same species. Skinks and other autarchoglossans, however, also obtain important information about their environment on the basis of chemical cues. Skinks are able to recognize potential mates and rivals, as well as food items

and predators, from pheromones or other substances produced by these animals. Chemical information is obtained through tongue-flicking; the autarchoglossan tongue is especially well designed to collect chemical cues and to deliver them to the vomeronasal organ, a chemical-sensing structure in the roof of the mouth.

Many lizards, especially males, bear glandular structures on scales of the thighs and/or around the vent; such glands reach the surface by way of prominent pores. These are found on representatives of all families except the skinks, chameleons, and anguimorphan lizards. While the function of these glands remains poorly understood, they appear to play a role in sexual behavior. Although much lizard communication is visual or chemical, a few lizard groups rely, in part, on vocalizations. The most spectacular voices are possessed by geckos, but representatives of nine other families have also been known to vocalize. At least in geckos and some lacertids, calls seem to have particular value in night-time communication, when vision is poor. Whereas some gecko species are mute, or produce only small squeaks, others are highly vocal. The large tokay *Gekko gecko* barks loudly when disturbed and also produces a low warning growl. Other geckos actually call to attract mates. The bell geckos of southern Africa (genus *Ptenopus*) are the most accomplished at this. The mechanism of vocalization involves the passage of

▼ *Two flying lizards face-off on a rainforest log in Borneo. The prominent dewlap on the throat of these lizards (genus* Draco) *is used, like a semaphore flag, in social interactions between mates and territorial rivals, but the code has not yet been cracked.*

Leo Meier/Australian Picture Library

▲ *Bedecked with horns, knobs, warts, and armored scales, the thorny devil or moloch is in appearance one of the most bizarre of the world's lizards. It inhabits arid regions of Australia where it feeds mainly on ants. The various protuberances are arranged in such a way that dew or rain flows gradually into the crevices, and is then channeled to the corners of the mouth.*

▶ *With enormous extended frills, wide open mouth, and loud angry hisses, a frill-necked lizard warns off an attacker. This spectacular lizard inhabits tropical woodlands across northern Australia.*

air through the larynx, which may be equipped with vocal cords. Like frogs, lizards produce recognizable calls specific to their own species.

THREE MAJOR LINEAGES

Modern lizards are representatives of three major lineages, the Iguania, Gekkota, and Autarchoglossa. Although each has a variety of morphological features common to its component families, the diversity in body form and biological characteristics within any one of the lineages is staggering.

THE IGUANIANS

The iguanians include the agamids, chameleons, and the pleurodont iguanians (in turn divided into eight major lineages), and probably diverged from other groups early in the history of lizards, although no fossils have been found from before the late Cretaceous. The iguanians are fully limbed and are visually oriented. They use their large tongues to capture prey and gather other food. Prey are usually ambushed rather than pursued. Crests, fans, and dewlaps are common in the group.

Agamids

The Agamidae is a group of about 350 species in 45 genera living throughout the warmer regions of the Old World (except Madagascar and most oceanic islands). In form they closely resemble the American iguanids, to which they are relatively closely related. One of the distinguishing characteristics of this family is the presence of acrodont teeth—these are unsocketed teeth borne on the rim of the jaws. Chameleons also have this type of dentition. Most other lizards have pleurodont teeth—also unsocketed, but borne on the inner face of the jaw bones. Agamids are diurnal (day-active), visually oriented lizards, and all but a few species are oviparous (the female produces eggs that hatch outside her body). Some species live on the ground, others are tree-dwellers, and many others make their home among rocks. All are fully limbed and lack fracture planes in the tail, so the tail is never shed.

The largest radiations of agamids occur in

Michael and Patricia Fogden

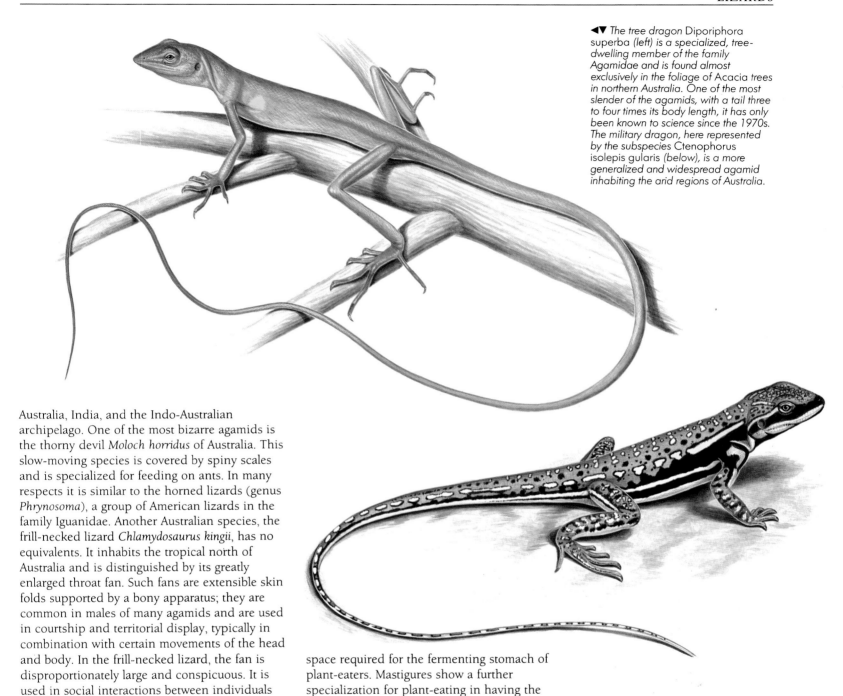

◄▼ *The tree dragon Diporiphora superba (left) is a specialized, tree-dwelling member of the family Agamidae and is found almost exclusively in the foliage of Acacia trees in northern Australia. One of the most slender of the agamids, with a tail three to four times its body length, it has only been known to science since the 1970s. The military dragon, here represented by the subspecies Ctenophorus isolepis gularis (below), is a more generalized and widespread agamid inhabiting the arid regions of Australia.*

Australia, India, and the Indo-Australian archipelago. One of the most bizarre agamids is the thorny devil *Moloch horridus* of Australia. This slow-moving species is covered by spiny scales and is specialized for feeding on ants. In many respects it is similar to the horned lizards (genus *Phrynosoma*), a group of American lizards in the family Iguanidae. Another Australian species, the frill-necked lizard *Chlamydosaurus kingii*, has no equivalents. It inhabits the tropical north of Australia and is distinguished by its greatly enlarged throat fan. Such fans are extensible skin folds supported by a bony apparatus; they are common in males of many agamids and are used in courtship and territorial display, typically in combination with certain movements of the head and body. In the frill-necked lizard, the fan is disproportionately large and conspicuous. It is used in social interactions between individuals and in displays to deter predators. The frill-necked lizard is one of several agamids whose powerful hind legs and long tail facilitate running bipedally—usually over short distances as part of a rapid escape behavior—but the frill-necked lizard may also stand on its hind legs and spread its magnificent frill in a defensive bluff attack.

Several agamids, such as the mastigures (genus *Uromastyx*), which are desert-dwelling lizards of western Asia and North Africa, reach large size. Mastigures are stocky, terrestrial lizards that use their short powerful legs to excavate burrows. They are strikingly similar in appearance to the chuckwalla *Sauromalus obesus*, an American desert iguanid. Mastigures and the chuckwalla are vegetarians. At least in iguanian lizards, large size seems often to be associated with the additional space required for the fermenting stomach of plant-eaters. Mastigures show a further specialization for plant-eating in having the blade-like cutting surfaces instead of more-typical teeth in the front of the jaw, whereas the back teeth are situated for crushing.

Another desert-dwelling group are the toad-headed agamids (genus *Phrynocephalus*), small lizards with fringed toes used for moving over sand, and for burial and excavation. They are insect-eaters and live in central Asia and the Middle East. In social interactions with one another, toad-heads coil the tail upward over the head. Although most toad-headed agamids are oviparous, a few species in mountainous regions retain eggs in the body until hatching, so that they give birth to live young.

The agamids of Asia include many tree-dwelling species of the forests. Most have relatively compressed bodies. One of these, the earless agama *Cophotis ceylanica,* is the only tropical agamid known to give live birth. Only about five young are born, in keeping with the relatively low number of eggs (1 to 27) typifying oviparous members of the family. This is a slow-moving, chameleon-like agamid with a strongly prehensile tail and a fleshy or scaly nasal ornament, which is believed to help it recognize individuals of its own species. Another Asian species, the water dragon *Hydrosaurus amboinensis,* is also semi-arboreal, but lives in close association with streams. It has a strongly compressed tail and toe-fringes that increase the surface area for swimming efficiency. The water dragon is the largest agamid, with a total length of over 1 meter (3¼ feet). The most specialized of the arboreal agamids inhabit the forests of Southeast Asia and peninsular India; the "flying dragons" (genus *Draco*) not only climb but glide through the air as well.

Relatively few species of agamids occur in Africa, but they include several tree-living species and numerous rock-living forms. The Namibian rock agama *Agama planiceps* is a specialist of boulder piles or rocky outcrops. Like many other agamids it is a sun-lover and spends much time basking and displaying to other members of its species. As in many agamids, there is marked color difference between the sexes (known as sexual dichromatism). Males are deep blue or purple with an orange or red head and tail. Females and juveniles have brownish or grayish bodies with yellow marks on the head and orange

shoulder spots. Color change is common in some species; for example, males of the Indian bloodsucker *Calotes versicolor* turn bright red following victory in battles with rivals.

Chameleons

The Chamaeleonidae is the most distinctive of all lizard groups. Nonetheless, chameleons have several anatomical and behavioral features that clearly link them to the Agamidae, from which they probably evolved; for example, like agamid lizards they have acrodont dentition—unsocketed teeth borne on the rim of the jaws. Chameleons are distributed throughout much of Africa (except the Sahara) and extend eastward to India and Sri Lanka and north to Spain. There are 135 species in six genera, all strikingly similar in body form. Ranging in size from tiny species of *Brookesia* less than 25 millimeters (1 inch) in total length, to giants of more than 550 millimeters (21½ inches) in total length like *Chamaeleo oustaleti.* All have laterally compressed bodies, prehensile tails, prominent, independently mobile eyes, and partially fused toes.

Another characteristic feature of chameleons is their projectile tongue, used to capture small prey which form the bulk of their diet. Insects and other arthropods are taken by most species, although small birds and mammals may be eaten by larger types. Each turret-like eye can be moved independently of the other, so a chameleon has the excellent depth perception necessary for

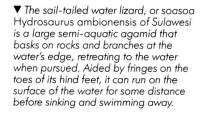

▼ *The sail-tailed water lizard, or soasoa* Hydrosaurus ambionensis *of Sulawesi is a large semi-aquatic agamid that basks on rocks and branches at the water's edge, retreating to the water when pursued. Aided by fringes on the toes of its hind feet, it can run on the surface of the water for some distance before sinking and swimming away.*

◀ A common or flap-necked chameleon Chamaeleo dilepis. The chameleons are predominantly an African group; their three most distinctive external characteristics are their prehensile tail, partly-fused opposable toes, and their extraordinary turreted, independently rotating eyes.

▼ Two Johnston's chameleons Chamaeleo johnstonii mating. The males of several species of chameleons have conspicuous horns, used in species recognition and perhaps in battles over mates or territories.

aiming its extremely long tongue and for judging distance in dense vegetation.

Most chameleons live in humid forest areas, and they are especially numerous in the rainforest belt of eastern Madagascar and highland forests of east Africa and the Cameroons. But they are also successful in Mediterranean climates and even in deserts. Their feet are equipped with opposable sets of partially fused toes, a condition called zygodactyly, which is an adaptation for grasping and is used by the tree-dwelling species for climbing. The limb girdles are also highly modified to permit the slender limbs to be held close to the body, as required by the small diameter of perches occupied by the lizard. Their prehensile tail acts as a fifth limb and unlike the tails of many other lizards, they are never shed. Although most chameleons are arboreal and move clumsily on the ground, arid-zone species such as the Namaqua chameleon Chamaeleo namaquensis spend most of their time on the ground, as do many of the stump-tailed chameleons (genus Brookesia) of Madagascar. The latter are tiny and drab in color, resembling dead leaves, and the tail is reduced to a small protuberance. Despite the apparent difficulty in walking on flat surfaces, both types of terrestrial chameleons retain the zygodactylous condition.

Males of virtually all species are territorial. Jackson's chameleon Chamaeleo jacksonii is one of many species in which males possess horns or other ornamentation on the head—devices that play a role in species recognition between the sexes and may be used in intense combat among the males. Breeding males of the South African dwarf chameleons (genus Bradypodion) may take on incredible colors when presented with another

► A tongue-lashing, chameleon-style. A Mediterranean chameleon Chamaeleo chamaeleon *scores another hit. The tip of the chameleon's remarkable tongue often exceeds 5 meters (about 16½ feet) per second as it speeds toward the unsuspecting fly.*

CHAMELEON TONGUES

Lizard tongues vary greatly in their structure, enabling scientists to use them as evidence of relationships between different groups. Those of autarchoglossan lizards (and snakes) are generally forked and often very slender, and are highly efficient as chemical-detection organs, collecting molecular cues about the environment. In geckos and flap-foot lizards the tongue is fleshier and serves in lapping up water or nectar; it is also used for "spectacle-wiping", cleaning the surface of the transparent scale over each eye. Some lizards, such as the Australian blue-tongued skink *Tiliqua scincoides*, have a strikingly colored tongue which

is used in defensive displays. Only in one lineage of lizards, the Iguania, does the tongue play an active role in grasping and bringing food to the mouth; in iguanids and agamids this ability is moderately well developed, but in chameleons it is developed to an amazing degree.

The tongue of all lizards is supported by the hyoid skeleton; the tip of the tongue is supported by one rod of the hyoid skeleton, the lingual process. Complex musculature moves the tongue as a whole, or certain parts of the tongue, as needed for swallowing, seizing prey, or other functions. In chameleons the tongue musculature

Stephen Dalton/NHPA

is exceedingly complex. During feeding, chameleons first judge the distance to their prey, then orient themselves to aim the tongue correctly. The mouth is opened and the hyoid apparatus is pulled forward, causing the tongue to protrude. An accelerator muscle, which extends the tongue out of the mouth, lies near the tip of the tongue. When activated it contracts, exerting a force around the lingual process of the hyoid. This serves to project the tongue tip out of the mouth (much as squeezing a wet bar of soap will cause it to shoot from your hand), bringing with it the retractor muscle, which at rest lies coiled around the more basal portion of the lingual process. The length of the retractor, usually equal to or greater than the head–body length of the chameleon, determines the distance that the tongue may be extended, but active firing of this muscle and/or the accelerator may allow the chameleon to extend the tongue a lesser distance. The outer surface of the tongue tip bears a glandular pad which contacts and grasps the prey by a combination of wet adhesion and muscular activity. Once the prey has been contacted the retractor muscle fires, pulling the entire apparatus back into the mouth. Finally, the entire hyoid is retracted, and the mouth closes. The tongue tip may exceed speeds of 5 meters (16½ feet) per second during the projection phase, requiring less than one-hundredth of a second for the tongue to reach the prey.

▲ *Chameleons are famed for their highly developed camouflage, but not all color changes are in response to the environment. The Knysna dwarf chameleon* Bradypodion damaranum *(top) is predominantly green at rest, but males develop the bright color pattern illustrated as a threat display to other males. The Malagasy chameleon* Furcifer lateralis *(above, right) is also usually dull green in color when resting among the foliage but takes on the striking disruptive pattern shown here when disturbed.*

male and engage in hissing, posturing, and biting. These species may breed several times in a year, giving birth to live young, a trait shared with several *Chamaeleo* species.

Most species lay eggs, however, with the number per clutch largely dependent on body size. One of the largest, Meller's chameleon *Chamaeleo melleri*, lays up to 70 eggs, and clutches of 30 to 40 are the rule for many species. Females usually bury the eggs in the ground or in rotting logs or other moist protected spots. During mating males often bite females. The hemipenes of many chameleons have highly complex, species-specific structures. Females can store viable sperm for long periods.

One well-known feature of chameleons is their ability to change color. This trait is shared by a variety of other lizards but is especially well developed in the subfamily Chamaeleonidae. The ability to blend into their surroundings, which is enhanced by their leaf-like shape, is useful in stalking prey. Color change is also associated with temperature and behavior. At night when it is cool, chameleons blanch and show up like white leaves on the ends of twigs. By sleeping in such spots the lizards are protected from predators such as snakes, who are too heavy for the slender branches. Another strategy, employed by smaller chameleons, is to drop to the ground and remain motionless when disturbed. Among the leaf-litter they are virtually impossible to see.

Pleurodont iguanians

Iguanid lizards are chiefly New World relatives of the agamids and chameleons. They are distinguished from these other iguanian groups by their pleurodont dentition—unsocketed teeth on the inner face of the jaw bones. More than 850 species in 54 genera are distributed from southwestern Canada to the southern tip of South America; another five species occur on the oceanic islands of Fiji and Tonga in the Pacific, and seven inhabit Madagascar, where agamids are absent. Iguanids are highly diverse in body form and structure and have one of the greatest size ranges among all lizards. Most lay eggs but several groups have evolved viviparity (giving birth to live young) independently of one another. All species are day-active.

Recent evidence shows that not all iguanids are each other's closest relatives, and eight major lineages of pleurodont iguanians (which may or may not be more closely related to each other than they are to agamids and chameleons) have been identified. Under one classification scheme, the eight groups are recognized as separate families—Iguanidae, Opluridae, Phrynosomatidae, Tropiduridae, Polychrotidae, Hoplocercidae, Crotaphytidae, and Corytophanidae.

Some of the most familiar pleurodont iguanians are the iguanids proper (also called iguanines). These include the green iguana *Iguana iguana* of Central and South America, which reaches a total length of more than 2 meters (6½ feet). A female can produce up to 71 eggs in a clutch. The green iguana is essentially arboreal, but it lays eggs on the ground and the hatchlings emerge from buried nests. The genus *Iguana* has close relatives, the terrestrial rhinoceros iguanas (genus *Cyclura*) of the Caribbean, but some of these have become extinct because of the actions of humans. The lizards in this group are vegetarian. (As described later, most other pleurodont iguanians, which are considerably smaller in size, are insect-eaters although many will also prey on other animals.) Closely related to the mainland and Caribbean iguanas are the iguanas of the Galapagos Islands, genera *Conolophus* and *Amblyrhynchus*. These giant lizards must have arrived on the Galapagos as a result of over-water rafting from South America, because the volcanic Galapagos have never been attached to another landmass. Charles Darwin and countless other visitors were impressed by the marine iguana *Amblyrhynchus cristatus,* a black-skinned lizard that feeds almost exclusively on marine algae, for which it dives. It is the only living lizard dependent on the marine environment. To rid itself of excess salt ingested when feeding, the marine iguana excretes concentrated salt crystals from a nasal gland. Temperature regulation is also a problem for this lizard. Although the Galapagos straddle the Equator, an undercurrent brings cold waters to the shore. When not

foraging, the marine iguana must spend much of the time basking on the rocks near the sea in order to raise their body temperature to the point where digestion can occur and further foraging can take place. The black coloration of the lizard facilitates rapid heat absorption. Thousands of individuals may bask at a given site.

· Banded iguanas (genus *Brachylophus*) occur in Fiji and Tonga. These too must have arrived over

▼ *Most lizards are carnivorous, feeding especially on insects. But in its diet of marine algae, the marine iguana* Amblyrhynchus cristatus *of the Galapagos islands presents a conspicuous exception. Here a male grazes in a tide pool.*

Mark Jones/AUSCAPE International

◄▼ *The marine iguana (left) is the only marine lizard. Normally black in color, for rapid re-warming after emergence from the cool water, this male is in full breeding dress. Like the marine iguana, the Fijian crested iguana* Brachylophus vitiensis *(below) probably evolved from mainland South American iguanas that accidentally drifted across the Pacific on floating vegetation.*

Michael and Patricia Fogden

▲ *A male ground anole* Anolis humilis *in a Central American rainforest. Male anoles have a conspicuous often brightly colored pouch, called a dewlap, on the throat, which is extended in display.*

water from the Americas, a journey of more than 9,000 kilometers (5,600 miles).

Anoles and their relatives (Polychrotidae) are primarily a South American and West Indian group, although one species, the green anole *Anolis carolinensis,* reaches well into the southeastern United States. Anoles are a huge group, accounting for 50 percent of all known iguanids. Their radiation has been particularly spectacular in the West Indies, where their ecology has been well studied. Anoles bear pads on their toes, similar to but slightly less well-developed than those of geckos. Most are effective climbers, but species in any one region tend to segregate according to their preferred perch heights, ensuring that food and other resources are available to each species. For example, in Puerto Rico, *Anolis cuvieri* occupies the crowns of trees, *A. evermanni* prefers tree trunks, *A. pulchellus* is active on bushes and in grass, and *A. cooki* may be found on the ground. Size, diet, shade tolerance,

and other features further segregate the anole species in any given area. Male anoles display using their brightly colored dewlap, and successful matings result in the production of a single egg. Many species have more than one clutch a year.

The sand lizards, horned lizards, fence lizards, and their allies constitute a particularly diverse group of iguanians, the Phrynosomatidae, widely distributed from extreme southern Canada to Panama. The most obviously distinctive genus is *Phrynosoma,* the horned lizards. These are flattened, with round bodies and spiny dorsal scales. They bear prominent backward pointing "horns" on the head, which are used in defense. Horned lizards are ant-feeding specialists, reminiscent of the thorny devil of Australia. Horned lizards dig burrows or cover themselves with sand. Four species give birth to live young, whereas others lay eggs. The short-horned lizard *P. hernandezi* occurs in shortgrass prairie, sagebrush country, and even open forests at moderate to high

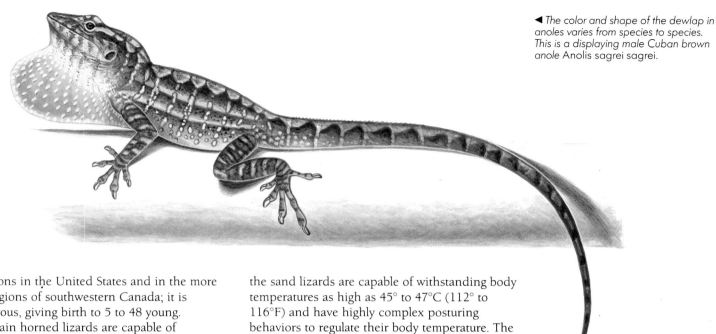

◄ The color and shape of the dewlap in anoles varies from species to species. This is a displaying male Cuban brown anole Anolis sagrei sagrei.

elevations in the United States and in the more arid regions of southwestern Canada; it is viviparous, giving birth to 5 to 48 young.

Certain horned lizards are capable of employing a rather specialized defense, in which blood is squirted from the eyes. This is accomplished by the lizard restricting the blood flow out of its head until mounting pressure ruptures delicate capillaries in the eyes. The thin stream of blood may travel up to 1.2 meters (4 feet) and may act as a deterrent if it contacts the eyes or mouth of a mammal predator.

Among several groups of sand lizards, Uma is the most specialized genus. These animals are active on the surface but bury themselves in sand dunes to escape predators and to sleep. Some of the sand lizards are capable of withstanding body temperatures as high as 45° to 47°C (112° to 116°F) and have highly complex posturing behaviors to regulate their body temperature. The zebra-tailed lizard Callisaurus draconoides shows sexual dichromatism (the males exhibiting bright blues patches on the belly and flanks that are absent in the females), and in most sand lizards pregnant females take on a color pattern indicative of their reproductive state. The zebra-tailed lizard is an especially fast runner and is often seen to run on its hind legs. It uses the black and white banded underside of the tail as a signaling device. The genus with the greatest number of species is Sceloporus, the fence lizards and their allies. These are a widely distributed

◄ The regal horned lizard Phrynosoma solare. Members of this North American group have evolved an exceptionally bizarre defense against predators: when under threat they can restrict blood flow from the head until mounting pressure ruptures small blood vessels in and around the eyes, resulting in a spurt of blood that may leap a meter (3¼ feet) or more.

group with broad ecological preferences. Some like the large *S. magister,* occur in rocky desert areas, while others, like the eastern fence lizard *S. undulatus,* favor forest edges and other sunny but not necessarily dry habitats. Typical for the family, eastern fence lizards are sexually dichromatic, with males having blue pigmentation on the sides and underside. These are accentuated during head-bobbing and other sexual and aggressive behaviors. Of about 75 species of fence lizards, 40 percent give birth to live young; this trait seems to have arisen several times within the genus *Sceloporus* and mostly characterizes those species that occur at high elevations.

Basilisks and their relatives (Corytophanidae) are a small group (three genera, nine species) of moderate to large forest-dwelling iguanids. Their range centers on Central America, extending north to Mexico and south to Ecuador. The best known

lizards in this group are the basilisks (genus *Basiliscus*). These are large green or brown lizards with prominent crests and dewlaps and long, powerful legs. The common basilisk *B. basiliscus* bears not only head ornamentation but also a sail-like crest on its back and tail. Basilisks can run on their hind legs, and may even run (for brief periods) on water. Their toes bear a fringe of scales on their lateral surface. At rest, this fringe folds over the axis of the toes, but in water the fringe is forced upward and serves to increase the surface area of the foot. When threatened or in search of food, a basilisk may run bipedally toward the water, where the combined effect of the fringes and its momentary high speed allows the animal to run on the surface of the water for a short distance before sinking. The tail is essential as a counterbalance in bipedalism, and basilisks are not able to shed their tail when caught by a

▶▼ *The South American equivalent of the Asian sail-tailed lizards, the basilisk or Jesus lizard* Basiliscus basiliscus *(right), also sports a large fin-like crest, and can run bipedally for a short distance across the surface of the water aided by fringes on its toes. The collared lizard* Crotaphytus collaris *(below, right) of arid southwestern North America can run for a distance on its hind legs at top speeds.*

predator, although most other New World iguanian lizards retain this ability. Basilisks are oviparous, but their close relative, the tree-dwelling helmeted iguanids (genus *Corytophanes*), are another of the many iguanian types that give birth to live young. Helmeted lizards are laterally compressed and have a large casque and crest on the head, which increases the apparent size of the head, thus serving as a deterrent to potential predators.

Crotaphytids are desert and plains-dwelling lizards of western North America. Two very closely related groups are included, the leopard lizards (genus *Gambelia*) and the collared lizards (*Crotaphytus*). They all have large heads and long limbs and tails and are exceptionally fast runners. The larger leopard lizards feed mainly on other reptiles, consuming a large number of smaller iguanid lizards.

Hoplocercids are lizards of tropical South America. The Brazilian spiny-tailed lizard *Hoplocercus spinosus* uses its tail as a defensive weapon when attacked. The spiny-tailed lizard is not truly a burrowing animal, but it digs shallow retreats in the soil and may use its tail to block the entrance and prevent a predator from reaching it. Spiny tails also occur in several other groups of iguanians, such as the tropidurids.

Tropidurids are a South American group, although some species live in the West Indies and the Galapagos. They include *Liolaemus magellanicus*, the southernmost lizard in the world, reaching Tierra del Fuego at the southern tip of South America. *Liolaemus* is a large genus including many high-altitude species in the Andes and its foothills. Like most other tropidurids, they are generally terrestrial lizards. Most members of this genus give birth to live young, in keeping with the harsh climate of their mountain home. Lava lizards (genus *Tropidurus*) are found on the mainland of South America and on the Galapagos, where their large numbers and bright colors make them highly conspicuous. Males engage in head-bobbing and push-ups; and confrontations, often involving displays rather than physical contact, are common. Size and color differences between the sexes are quite obvious in members of this genus.

On the island of Madagascar there are seven species of pleurodont iguanians (oplurids). These are lizards of moderate size, generally sun-loving, and all are egg-laying. Six species belong to genus *Oplurus* (two are tree-dwellers, and four live among rocks). The other lizard, *Chalarodon madagascariensis,* inhabits sandy areas where it makes burrows. Males in combat may be especially violent, and combat may involve acrobatic twists in the air while engaged in head biting. The presence of these lizards in Madagascar remains a biogeographic mystery to scientists but certainly seems to date from an early period in the history of the Iguania as a whole.

Michael and Patricia Fogden

THE GEKKOTANS

Only three living families, the Eublepharidae, the Gekkonidae, and Pygopodidae, constitute the gekkotan lineage. The group is an ancient one, as evidenced by *Eichstaettisaurus*, a fossil gekkotan from the Jurassic, 200 to 145 million years ago, which bears some striking similarities to living forms. Although differing greatly in body form, the fully limbed geckos (Eublepharidae and Gekkonidae) and the reduced-limbed flap-footed lizards (Pygopodidae) share a number of important features. The absence of temporal arches lends a lightly built appearance to the skull and renders it especially mobile. Males of most species have cloacal bones—paired structures embedded below the skin on either side of the vent. Both males and females have cloacal sacs, whose function we do not yet know. Females generally lay two eggs, although one or very rarely three eggs may be laid. Gekkotans are mostly nocturnal, and the replacement of moveable eyelids by a fixed transparent spectacle characterizes most of the group. The eyes are large and the pupils of nocturnal geckos dilate greatly at night to let in sufficient light to allow activity. In bright light the pupils close to a vertical slit or a series of pinholes.

▲ *A high-casqued lizard* Corytophanes cristatus *in threat display, Costa Rica. Inhabiting mainly rainforests, they have a characteristic habit of perching head-downward on tree trunks.*

David Hughes/Bruce Coleman Limited

▲ *The Namib web-footed gecko*
Palmatogecko rangei *of Africa. With large eyes, geckos as a group are characterized mainly by their adaptations for night-time hunting by sight.*

▼ *Recent studies of the leopard gecko* Eublepharus macularius *have shown that the sex of their offspring is determined by the temperature at which the eggs are incubated.*

Geckos

Next to the skinks, geckos (Eublepharidae and Gekkonidae) are the most numerically diverse group of lizards. More than 900 species in 90 genera are known, and they occupy all continents except Antarctica. Although they are especially numerous in the tropics and subtropics, some species live in cooler areas from northern Italy to southern New Zealand, and a few have even adapted to harsh alpine conditions. Most geckos

are small lizards, reaching a total length of no more than 15 centimeters (less than 6 inches), but a few exceed 30 centimeters (1 foot), and one species, the giant New Zealand gecko *Hoplodactylus delcourti,* which until recently inhabited New Zealand, was over 60 centimeters (2 feet) long. The largest living gecko, Leach's giant gecko *Rhacodactylus leachianus,* lives in the rainforests of New Caledonia and reaches 40 centimeters (15¾ inches) total length and may weigh 600 grams (1½ pounds).

One possible reason for the success of the geckos has been their nocturnal habits. The majority of geckos are active at night or are crepuscular (active at twilight and before sunrise). No other large group of lizards has specialized to such a degree in night-time activity. Geckos are therefore able to take advantage of the lack of competition from other lizards during the hours when a huge number of insects and spiders are available as prey.

Two families of geckos are recognized. Members of the Eublepharidae retain many primitive characteristics. Most obviously, they have moveable eyelids and their feet do not have the complex climbing system typical of many other geckos (see page 151). All species lay two eggs with a leathery shell, and all are primarily nocturnal. In at least some species, including the leopard gecko *Eublepharis macularius* of southwest Asia, the sex of an individual is determined by the temperature experienced by the embryo in its egg, a condition known chiefly in turtles and crocodilians. At present this group comprises only 25 species occurring in Africa, Asia, and North and Central America.

One species, the banded gecko *Coleonyx variegatus,* is a small, thin-skinned resident of the arid southwest of the United States and adjacent areas of Mexico, in a wide variety of microhabitat types, and it is an insect-eater. The banded gecko is inactive for as much as half of the year, when night temperatures fall below about 23°C (73°F).

The male uses visual and chemical signals to locate and identify a female. Courtship involves tail flicks by the male, and he ultimately mounts the female, biting her nape, and pinning her body down, while maneuvering his tail under hers to reach her cloaca with his hemipenis.

All non-eublepharid geckos (Gekkonidae) and their close relatives, the flap-footed lizards (Pygopodidae), lack eyelids. Instead the eye is covered by a transparent scale, called the spectacle, which is cleaned by periodic licking. (Similar transparent windows in the eyelids have evolved in night lizards as well as some skinks and lacertids.)

Many geckos are terrestrial and live in deserts or other arid habitats. However, perhaps the feature most closely associated with geckos is their ability to climb, even on smooth surfaces such as glass. This feature has certainly played a major role in the diversification of geckos and their success throughout the world. Climbing ability is dependent upon claws and/or scansors, the adhesive pads on the toes. In many instances the type of toe pad is revealed by the name of the genus, and each type seems to have particular advantages on certain types of surfaces.

The Diplodactylinae is a subfamily of geckos restricted to New Zealand, where they are the only gecko group represented, and Australia and New Caledonia where they form the bulk of the gecko fauna. It is probable that the flap-footed lizards, which are also restricted to the Australian region, evolved from within this group of geckos and both are sometimes placed in a separate family, the Diplodactylidae. Most diplodactylines lay eggs, but the geckos of New Zealand (*Naultinus* and *Hoplodactylus*) and a single New Caledonian species, the rough-snouted giant gecko *Rhacodactylus trachyrhynchus*, give birth to live young. The evolution of this feature has not been fully explained, but in the New Zealand geckos it was associated with cool temperatures experienced by the animals and the need to regulate the

temperature of the embryos. Gestation may take almost a year, and like other geckos, only two young are produced in each litter. The most successful genus in the group is *Diplodactylus*: 40 species, generally only 45 to 85 millimeters (1¾ to 3¼ inches) head–body length, are distributed across mainland Australia. The name of the genus comes from the divided scansors or toe pads that characterize the group. Although many *Diplodactylus* species are terrestrial, most are capable of climbing as well. Certain members of this genus, such as the western spiny-tailed gecko *Diplodactylus* (or *Strophurus*) *spinigerus*, have large glandular structures in their tails which produce a sticky substance that is squirted from the tail up to a meter (and with some accuracy) by muscular contraction. Although the secretion is not toxic, it may be distasteful to predators or may foul the feeding apparatus of large insects and spiders that prey on these geckos. One of the oddest

Hal Cogger/Nature Focus

▲ Virtually all geckos lack eyelids, the eye being instead protected by a fixed transparent scale as in this rainforest gecko Rhacodactylus chahoua *from New Caledonia. This species also has the moss- and lichen-like protective coloration common in many arboreal rainforest lizards, making it perfectly camouflaged when foraging at night on the trunks of large forest trees.*

▼ Velvet geckos possess tiny, even scales which give their skin a velvety texture, hence the common name. The genus Oedura *is endemic to Australia; the southern spotted velvet gecko* O. tryoni *shown here is found on granite outcrops on the east coast.*

Australian diplodactylines is the spiny knob-tailed gecko *Nephrurus asper,* whose tail is short and ends in a small knob, like the other members of its genus. The function of the knob is unclear, but it may play a sensory role. These geckos spend the daylight hours in burrows and emerge at night to feed on insects, spiders, and other lizards. By far the most numerous and widely distributed geckos are those of the subfamily Gekkoninae. The species are classified in almost 60 genera and, with the exception of New Zealand and parts of the United States, they reach to the limits of the family's distribution. Some of the largest genera are each characterized by a different type of toe structure, which has apparently contributed to their success in adapting to different habitats.

In Africa, south of the Sahara, members of the genus *Pachydactylus* are well represented. These are mostly terrestrial forms that attain their greatest diversity in the rocky arid regions of Namibia and South Africa. As many as eight species may live in the same area, partitioning the available resources. From this genus several others have evolved. Most spectacular of these is the Namib web-footed gecko *Palmatogecko rangei,* which has translucent skin, extremely thin legs, and large webbed feet used for digging burrows in the sands of the coastal Namib Desert. The webs are fleshy, but contain small cartilages that support a delicate system of muscles which coordinate the fine sand-scooping motions of the feet. Adults frequently feed on small spiders, but larger white lady spiders in turn regularly capture and eat the smaller geckos.

In North Africa *Tarentola* is the dominant gecko genus. It also extends to southern Europe, where *T. mauritanica* is a common house gecko. Like a number of geckos throughout the world, this species has adapted well to the presence of humans and thrives on the walls of buildings where lights attract suitable insect prey.

A few species of geckos have been successful in colonizing vast geographic areas, being transported accidentally with humans and their possessions. The house gecko *Hemidactylus frenatus,* for example, ranges from India across Southeast Asia and far into the Pacific. It continues to spread to islands in the central Pacific, sometimes causing local extinctions of other gecko species. One of the features that makes these animals successful in this regard is their hard-shelled egg. Most reptiles have egg

▶ *Several geckos have unusually shaped tails, but the knob-tailed gecko Nephrurus asper has one of the oddest: as its name suggests, the tail is short and ends in a small knob of uncertain function. One of Australia's largest geckos, this species lives in arid, rocky environments across the northern half of the continent.*

Jean-Paul Ferrero/AUSCAPE International

shells that are leathery, but gekkonine geckos have hard-shelled eggs which are highly resistant to desiccation and can even survive prolonged immersion in salt water.

The genus *Hemidactylus* is one of the most successful of all gecko genera, with 75 species ranging throughout Africa (a great diversity in Ethiopia and Somalia), Asia (many Indian forms), islands of the Pacific, and South America. A similarly successful group (85 species) are the bent-toed geckos, *Cyrtodactylus, Cyrtopodion,* and their allies. Although formerly regarded as a single genus, several lineages of bent-toed geckos are now recognized. These are rock-, ground-, or tree-dwelling species with reduced climbing pads on their toes. The claws are large and prominent, however, and most are efficient climbers. Bent-toed geckos are most diverse in Asia, extending from the Mediterranean, through the arid and mountainous regions of western Asia to tropical Southeast Asia, the Indo-Australian archipelago, and out into the islands of the Pacific. The mourning gecko *Lepidodactylus lugubris* is a particularly widespread species in the Pacific. In addition to having a calcareous eggshell, its spread has probably been enhanced by parthenogenesis, the ability of a female to reproduce without the need for fertilization of the eggs and hence without the need for a male.

The vast majority of geckos are insectivorous, attacking any arthropods small enough to ingest. Some of the larger geckos, however, may take vertebrate prey such as small mammals, birds, and especially other lizards. The tokay *Gekko gecko* is a good example. It has even been known to attack snakes. Tokays are also especially vocal lizards and growl or bark loudly in social and defensive behaviors. The genus *Gekko* is one of several highly successful genera with their center of distribution in tropical Asia. Gecko colors are usually drab browns or grays, but a few species, such as tokays which are bluish with orange spots, deviate from this pattern. By far the most colorful species are the day geckos (genus *Phelsuma*) of Madagascar and the Indian Ocean islands. In addition to taking arthropod prey such as insects and spiders, these bright green, red, and blue lizards seek out flowers and eat nectar and pollen. This habit may, in fact, be widespread among geckos. (It has been reported in the New Zealand diplodactyline geckos of the genera *Hoplodactylus* and *Naultinus*.) Day geckos have greatly reduced claws and rely entirely on their toe pads when climbing.

The sphaerodactyls are an American offshoot of the Gekkoninae. The ancestors of these lizards probably came from Africa, either traveling on the South American continent as it drifted

▲ *The blue-tailed day gecko* Phelsuma cepediana *(top) is a colorful gecko from the islands of Réunion and Mauritius. Being active during daytime and a tree-dweller, it has round pupils and greatly reduced claws on its digits relying solely on its toe pads when climbing. The ring-tailed gecko* Cyrtodactylus louisiadensis *(above) is a widespread species of the Australasian region and is the largest gecko on the Australian continent. For an obvious reason the common name of this genus is bent-toed geckos.*

G.I. Bernard/NHPA

significant predators. Some species have very fragile skin which may be an adaptation to escape such attackers. Similarly weak skin is also found in some gekkonines, especially island species. Although many species of sphaerodactylines are diurnal (day-active), they are secretive and active in protected microhabitats. The white-throated gecko *Gonatodes albogularis* and its allies are generally the most conspicuously day-active sphaerodactyls. Sexual dichromatism is seen in this species with males exhibiting a bright yellow head, whereas females are more subtly patterned; sexual dimorphism, such as this, is rare in geckos and is found in only a few groups, most of them diurnal.

Pygopodids (flap-foots)

The flap-footed lizards of the family Pygopodidae have extremely long bodies and reduced limbs, so that they look remarkably like snakes. During evolution, forelimbs have been lost entirely, whereas hind limbs are invariably just small, flattened flaps lying close to the cloaca. Though small, the limbs may be used in some types of locomotion (through vegetation) and in courtship and defensive behaviors. Most of the 36 species of pygopodids are endemic to Australia—one species is restricted to New Guinea, and another occurs on both landmasses. Despite their striking dissimilarity in overall appearance, flap-footed lizards are closely related to geckos, and may in fact be an offshoot of a particular group of geckos endemic to Australia and the southwest Pacific, the diplodactylines. Like most geckos they do not have eyelids. Also, in common with the majority of geckos, all pygopodids lay two eggs per clutch.

Some pygopodids, such as the species of *Aprasia*, are chiefly burrowers that feed on insects, whereas others have become specialized for life above ground and they retreat to the burrows of spiders or other animals or into the dense clumps of spinifex grass found in arid Australia. Members of the genus *Delma* are slender, surface lizards and employ a unique jumping mechanism in their repertoire of defenses. When agitated they use their long tail to generate an upward thrust and

▲ *Many nocturnal geckos rely heavily on crypsis, with subtle and intricate patterns of sober shades of gray, brown, and green to protect them by day. But day-active species, like this multicolored gecko* Gonatodes ceciliae *of Trinidad, are often brilliantly colored and boldly patterned.*

▼ *The yellow-headed gecko* Gonatodes albogularis fuscus *is a member of the subfamily containing the smallest geckos, the Sphaerodactylinae. One of the few sexually dimorphic geckos, only the males display the yellow color on the head.*

westward in the Cretaceous period 145 to 65 million years ago or rafting across the narrow and newly formed Atlantic Ocean shortly thereafter. Their closest relatives are probably among the semaphore geckos (*Pristurus*) of the Middle East, or *Quedenfeldtia*, a small North African form. These are mostly tiny geckos, including the smallest living lizard, the Monito gecko *Sphaerodactylus parthenopion* of the Virgin Islands, which reaches only 17 millimeters (⅔ inch) head–body length. Most of the 85 species of the genus *Sphaerodactylus* are island-dwellers, living among fallen palm fronds or other litter in forests or open areas in the West Indies. A few, such as the Cuban dwarf gecko *S. bromeliarum*, occupy bromeliads and leaf axils in trees. The Americas are generally poor in geckos, and members of this genus account for the majority of all American gekkonids. These lizards are so small that arthropods such as spiders may be their most

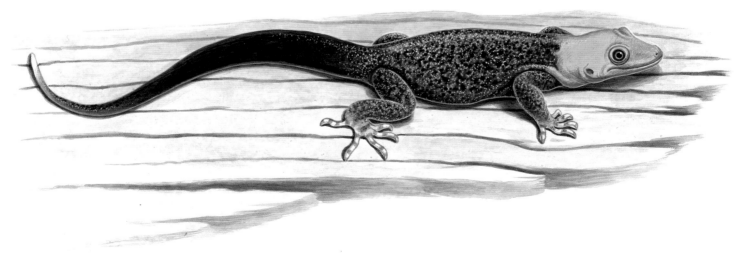

STROLLING ON THE CEILING

Many lizards climb, but only a few have evolved specialized structures that enable them to scale smooth surfaces without the aid of claws. Such climbing devices are found in many representatives of the family Gekkonidae, a few skinks, and the polychrotid lizards of the genus *Anolis*. The toes of climbing geckos bear enlarged overlapping plates called lamellae on their undersurfaces. Each lamella is covered by a field of setae, which are microscopic projections of the skin; these tiny spatula-shaped prongs vary in length from species to species, but in most geckos they measure 10 to 100 micrometers. When the gecko climbs, the expanded tips of the setae are brought into very close contact with the surface, and weak forces between the molecules form a temporary bond between the two surfaces. Although the forces acting on any one seta are minute, there may be more than one million setae per toe, so the additive adhesive force may be quite impressive. In order to ensure that as many of the setae as possible actually come into contact with the surface, geckos also have a complex internal mechanism within the toe.

A lamella is the most external portion of a structure called a scansor. Inside the scansor there is an extensive network of blood vessels. These in turn are linked to a sinus, a small blood reservoir beneath the bones of the toe. The gecko can shut off this part of the circulatory system from the rest of the body by a series of valves; and when the animal pushes onto the bones of the toe above the sinus, it pressurizes the network of vessels, causing them to expand and push onto adjacent lamellae, forcing them onto the substrate. In this way the expanded toe pad is made to conform closely to any irregularities in the substrate, maximizing the number of setae that can be involved in adhesion. Tendons in the scansors also permit fine control so that each scansor can be moved independently of the others.

The mechanism of adhesion poses a special problem when the gecko is walking. In order to lift the foot a gecko must depressurize its blood sinus and the network of blood vessels and break the weak bonds that hold the setae to the surface. This is accomplished by rolling the toes up from tip toward the base, thus forcing blood back toward the foot and peeling the setae away from the surface. All of this happens with every single step the gecko takes!

The origin of the whole adhesive complex is obscure, but the setae themselves appear to be elaborations of surface features that are present on the skins of all lizards. The climbing ability of geckos has certainly contributed to their success in habitats in trees and rocks. This form of adhesion is perhaps most useful on hard, smooth surfaces to which lizards with claws cannot cling. Paradoxically, surfaces such as glass, which

Hal Cogger/Nature Focus

▲ *Geckos are a familiar part of the normal household fauna in dwellings throughout the tropics and subtropics, especially in village homes that incorporate natural building materials from surrounding forests. Mainly nocturnal, these geckos scuttle freely over walls and ceilings as they hunt moths and other insects. This is a stump-toed gecko* Gehyra mutilata *from the Cocos (Keeling) Islands in the Indian Ocean, where it is being displaced gradually by more aggressive recently introduced species.*

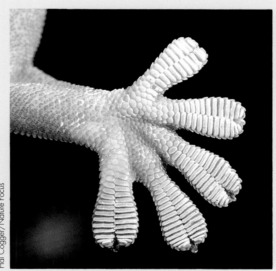

Hal Cogger/Nature Focus

◀ *The foot of a giant cave gecko* Pseudothecadactylus lindneri *from Australia's Arnhem Land, showing the lamellae on the underside of the toes.*

intuitively seem highly challenging for a climbing animal, are quite easily negotiated by geckos. The basis for the molecular bonds between the substrate and the setae is the amount of adhesive force of the surface. This is known as surface energy. It can be measured in relative terms by putting a drop of water on to a substrate; if the drop spreads out, the surface has a high degree of attraction for the water molecules and thus high surface energy; if the drop remains beaded-up, the attractive force is small and surface energy is low. Thus glass, with reasonably high surface energy is no challenge for a gecko, but the waxy surface of a plant leaf may be.

The adhesive mechanisms of the other climbing lizards have not been investigated in such detail but do not seem to be so complex. Nor are these lizards as accomplished climbers as geckos.

lift the entire body from the ground. Several jumps may be completed in succession, with the animal changing direction erratically each time. *Delma* species, like other pygopodids, are capable of vocalization, and this may also be used as part of the defensive strategy of the lizard. Burton's snake-lizard *Lialis burtonis*, with its characteristic elongate snout, is the largest Australian pygopodid at 59 centimeters (23¼ inches) total length, and feeds exclusively on lizards and snakes. It has pointed teeth that are hinged at their bases, which aid in capturing its prey; the smooth-scaled skinks that constitute much of the diet easily slide past the "folded teeth", but if they attempt to move backward, out of the snake-lizard's mouth, the teeth are pushed into their erect posture preventing escape. This species exhibits amazing variability in color and pattern. It is also the most widespread pygopodid, ranging across all of Australia except the extreme southwest and southeast, and occurs also in New Guinea. Another widespread flap-footed lizard, the black-headed scaly-foot *Pygopus nigriceps*, is characterized by a black crown and nape and closely resembles a venomous elapid snake that occurs in the same area. On this basis it has been regarded as a potential mimic of the snake.

THE AUTARCHOGLOSSANS

The third major lineage of lizards is the Autarchoglossa, which contains the twelve remaining lizard families. These lizards are mostly terrestrial, burrowing, or living among rocks. Few generalities can be applied to the group as a whole, but many species have osteoderms (bony plates) in the skin, and most rely heavily on chemical cues in their environment.

Autarchoglossa is further divided into two subgroups: the Anguimorpha (five families) and the Scincomorpha (seven families).

GROUP ONE: THE ANGUIMORPHS

The five types of anguimorph lizards are the anguids, beaded lizards (helodermatids), the Bornean earless monitor, true monitors (including goannas), and knob-scaled lizards. Although relationships are unclear, snakes share a number of features with some of the anguimorph lizards and some scientists believe it may be this group from which they evolved.

Anguids

The family Anguidae is a moderately diverse group of about 100 species, most of which inhabit the Northern Hemisphere. The family includes fully limbed and limbless species. Their scales contain bony osteoderms that give the lizards a hard, rigid feel. In two of the four subfamilies, Gerrhonotinae and Anguinae, the large rectangular scales of the back give way to small granular scales on the underside of the body. These cover an osteoderm-free fold of skin that runs most of the length of the body.

Alligator lizards and their allies are members of the subfamily Gerrhonotinae. This group extends from southwestern Canada to the tropics of Central America and is especially diverse in Guatemala and Mexico. About 40 species are known, and almost half of them belong to the tree-dwelling genus *Abronia*. Most species in this genus occur in forests in mountainous regions. Other alligator lizards are terrestrial or semi-burrowing and occupy a wide range of habitats from desert mountains to tropical lowlands.

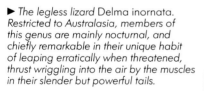

▶ *The legless lizard* Delma inornata. *Restricted to Australasia, members of this genus are mainly nocturnal, and chiefly remarkable in their unique habit of leaping erratically when threatened, thrust wriggling into the air by the muscles in their slender but powerful tails.*

Esther Beaton/AUSCAPE International

◀ Many alligator lizards (genus Abronia) inhabit rainforests of Central and South America, some species reaching 30 centimeters (11 inches) or more in length. Members of this genus have prehensile tails.

George Bernard/NHPA

The southern alligator lizard *Elgaria multicarinata*, and the northern alligator lizard *E. coerulea* occur together in much of western North America. The two species are similar in appearance, but differ in reproductive mode: the southern alligator lizard is oviparous, laying 6 to 17 eggs in burrows or other protected sites in the oak-grasslands it inhabits; the northern alligator lizard, which is somewhat smaller, gives birth to 2 to 15 live young about three months after mating. The northern alligator lizard is especially adapted to colder conditions and tends to occupy the margins of coniferous forests or other slightly cooler and moister sites than the southern

alligator lizard. Both species are typical of terrestrial members of the subfamily, in that they are generally secretive and feed on insects and spiders, small mammals, birds, and reptiles. *Elgaria* species can bite fiercely in defense but do not appear to be territorial. Although chiefly ground-dwelling, they also climb with the aid of a partially prehensile tail. The largest member of the subfamily is the Texas alligator lizard *Gerrhonotus liocephalus*, which reaches 20 centimeters (8 inches) head–body length. Like most members of the group, it is widely but incorrectly believed by many local people to be venomous.

Most members of the second anguid subfamily,

153

the birds themselves are eaten (the latter as carrion), as are the feces of the birds. Stranger still, galliwasps mob adult boobies returning to the nest with food intended for their chicks, causing the birds to regurgitate prematurely and thereby provide the lizards with food.

There is a tendency for the limbs to be reduced in some galliwasps. This is most accentuated in the South American worm lizards, genus *Ophiodes*, a small group of diploglossines retaining only rudimentary hind limbs. They are endemic to mainland South America and are strikingly similar to the anguines.

The anguines—the slow worm and glass lizards—are the most geographically widespread of the anguid subfamilies and the only group to occur in the Old World. They are characterized by the absence of limbs. The slow worm *Anguis fragilis*, found throughout much of Europe and Britain, is a highly adaptable lizard and survives today even within the confines of the City of London, occupying the railway right-of-ways that cross the city. The slow worm gives birth to 3 to 26 live young after about three months of gestation, although this is variable and if conditions are unfavorable the young may remain in the uterus of the mother.

The largest genus in the subfamily is *Ophisaurus*, with 12 species. Its representatives occur in North America, southern Europe, across Asia to Borneo, and marginally in North Africa. Some recent fossil and phylogenetic studies seem to support the hypothesis that the Old World and New World species had separate origins and evolved limblessness independently. All members of the genus are oviparous and the female remains with the eggs during incubation, although she does not actively defend the nest. The scheltopusik *Ophisaurus apodus*, of southeastern Europe and southwestern Asia, is the largest member of the genus and of the entire anguid family; large individuals may reach 1.4 meters (4½ feet) in total length. This species feeds on small mammals, snails, insects, and other reptiles. Smaller species in the genus are primarily insectivorous but may take nestling mammals and small amphibians and reptiles too. The tails of these lizards are extremely long—two and a half times the body length in one subspecies of *O. attenuatus*, the slender glass lizard of North America. The "glass lizard" name has been given to these animals because of their tendency to "break" or autotomize the tail. The slender glass lizard occurs in tallgrass prairie habitats, where it "swims" through the grass in search of prey or hides under vegetation or in burrows co-opted from other animals. These burrows are used for hibernation, as nests, and as refuges from predators and inclement weather. In cool weather the slender glass lizard is active during the day, but as the weather becomes warmer it prefers the

Aldo Brando Leon/Oxford Scientific Films

▲ *The subfamily Diploglossinae, or galliwasps, has a distribution centered mainly on the islands of the Caribbean. This is the white-spotted, or Malpelo galliwasp Diploglossus millepunctatus, restricted to Malpelo island off the coast of Colombia.*

Diploglossinae, are known as galliwasps. There are about 40 species recognized at present. They are skink-like animals with smooth overlapping scales and relatively short legs. Although there are some mainland species, most galliwasps are found on the islands of the Caribbean. Like the alligator lizards this group also contains both egg-laying and live-bearing forms. As few as one or as many as 34 offspring may result from galliwasp clutches or litters. Most galliwasps are primarily terrestrial or semi-burrowing, although many species often climb trees and other vegetation. Both nocturnal and diurnal species exist; a couple of the largest species, *Diploglossus anelpistus* and *D. warreni,* at about 28 centimeters (11 inches) head–body length, are night-active. Although galliwasps live in many types of habitats at all elevations up to 2,500 meters (7,600 feet), most prefer somewhat moister microhabitats. The Malpelo galliwasp *D. millepunctatus,* endemic to rocky Malpelo Island off the Pacific coast of Colombia, is a relatively large galliwasp, growing to at least 25 centimeters (10 inches) head–body length. It has some of the strangest feeding habits of any lizard. Malpelo is a small island, and resources are limited; few potential prey items of an appropriate size for the galliwasp can be found. Crabs and seabirds (mostly boobies) are plentiful, however, and provide most of the lizard's diet. The crabs and twilight and before dawn or, in especially hot periods, the night-time. During winter months the lizards hibernate in deep tunnels of rodents or moles. The slender glass lizard mates in late spring, and 5 to 16 eggs are laid in early summer, hatching by late summer or early fall. Females breed only every second year, as they lose significant weight while egg tending and must renew energy reserves the year after laying. As might be predicted from the low reproductive output, glass lizards are relatively long-lived, reaching at least 10 years of age for American species and perhaps 20 for the scheltopusik.

There are only two species in the anguid subfamily Anniellinae. They are the California limbless lizards (genus *Anniella*). Both are small (10 to 16 centimeters, or 4 to 6¼ inches, head–body length), limbless, and burrowing, and they inhabit coastal and inland regions of California and Baja California. They are regarded by some herpetologists as representing a separate family, the Anniellidae, closely related to anguids. Anniellids lack external ear openings but have unreduced eyes with moveable lids. The California limbless lizard *Anniella pulchra* occurs in California, where it lives in sandy soil around the bases of bushes; coastal sand dunes are favored sites, as are regions of alluvial soils in the inland where the substrate is suitable for the lateral undulations used by the lizard in sand-swimming. Like most anguids it is primarily insectivorous, catching adult and larval insects near the surface during the late afternoon or evening. California legless lizards are viviparous, giving birth to small litters of one to four young, which are relatively large when born in the fall. Relatively cool temperatures and access to soil moisture are favored by anniellids.

▼ *The slow worm* Anguis fragilis *is one of the hardiest and most adaptable of European lizards. Long, slender, and snake-like—a resemblance extending even to a forked tongue—it is often common even at the heart of major cities. It gives birth to living young after an extremely variable gestation period. Remarkably long-lived for such a small creature, some captive individuals have been known to live for 20 years or more.*

Beaded lizards

The family Helodermatidae contains only two living species: the Mexican beaded lizard *Heloderma horridum,* which occurs along the Pacific coast of Mexico and Guatemala; and the gila monster *H. suspectum,* from the southwestern United States to northwestern Mexico. These are the only venomous lizards in the world. They are most closely related to the true monitor lizards and the Bornean earless monitor, although Helodermatidae has existed as a recognizable group for at least 65 million years, since the late Cretaceous, and a number of species occurred in the Tertiary of North America. Both species are large—the head–body length of the Mexican beaded lizard can be 52 centimeters (20½ inches), or total length 1 meter (3¼ feet), and the gila monster's head–body length can reach 33 centimeters (13 inches)—and slow-moving. The scales are small, bead-like, and do not overlap, and all but some of those on the underside are underlaid by bony osteoderms. The two species are generally similar in appearance. They are stocky with broad heads and are typically black with varying amounts of yellow or pinkish spotting or reticulations. The tails are short in both species, especially so in the gila monster, and are used for fat storage.

Beaded lizards spend much of their time in burrows and generally avoid extremely high temperatures. Only in the cooler spring months and overcast days are they active during the day. They forage on the ground but may climb trees in search of prey as well, relying largely on chemical cues, such as smell. This is especially true of the more slender-bodied Mexican beaded lizard, which occupies not only arid regions but also tropical deciduous forests. Gila monsters also inhabit a variety of habitats but are most often associated with rocky slopes in areas of desert scrub, grassland, or oak woods. Both species feed on a variety of prey but rely heavily on the nest young of rodents and other small mammals, and bird and reptile eggs. Mammalian prey, detected through smell and sound, are usually dug out of the ground with the powerful forelimbs. The gila monster mates in late spring, often after intense combat between rival males. As many as 12 eggs are laid by the female and require about ten months before hatching.

Earless monitor lizard

The family Lanthanotidae contains only one species, the Bornean earless monitor *Lanthanotus borneensis.* It is closely related to the true monitors (family Varanidae) and is also allied to the beaded

VENOMOUS LIZARDS

Beaded lizards (family Helodermatidae) are the only lizards to possess venom. Evidence for the existence of a venom-delivery system is present even in an early member of the family, *Paraderma bogerti,* which lived more than 65 million years ago. Unlike venomous snakes, the lizards have venom glands in the lower jaw, which bulges conspicuously. Individual ducts lead from the glands to each of the lower teeth, which have grooves in the front for carrying venom. Venom is delivered with every bite, but there is no forceful ejection from the glands, and the lizard must chew the venom into its victim. The venom is generally not needed to subdue prey, as these lizards often feed on eggs or relatively defenseless prey (young mammals), so although venom may have evolved in association with feeding it is now used mainly as defense for these slow-moving lizards, which may be exposed for long periods while foraging in open habitat. Beaded lizards are immune to their own venom, which may be injected during territorial combat between males. The striking black and pink coloration of the gila monster *Heloderma suspectum* may serve either as camouflage or as a warning to potential predators.

The bite of the gila monster results in localized swelling and severe pain and may cause vomiting

John Cancalosi/AUSCAPE International

or faintness but is usually not fatal for humans. The mechanical damage done by the teeth and powerful jaws may also be significant. A gila monster may maintain a defensive bite for many minutes, during which the venom will continue to be released.

▲ *The gila monster inhabits arid regions of Mexico and the American southwest.*

Philippa Scott/NHPA

lizards, to which it bears some resemblance. The earless monitor is dark brown in color and grows to about 20 centimeters (8 inches) head–body length. The body scales are small, but several rows of enlarged tubercles run down its back. It is earless only in that it lacks a tympanum, or external eardrum, but it is quite capable of detecting sounds. This poorly known nocturnal lizard lives in the northern part of the island of Borneo, where it burrows and swims in search of prey, including earthworms. It may be able to climb as well, but is slow-moving on land. Its limbs are small, and burrowing is accomplished mainly by movements of the head; in water it moves by lateral undulations of the body. Like its closest relatives, the Bornean earless monitor is oviparous (egg-laying). Most details of its physiology, behavior, and ecology remain unknown.

Monitor lizards

The family Varanidae includes the monitors or goannas. All 48 species are classified in the genus *Varanus,* although some researchers recognize more species or divide the genus into subgenera, or species groups. All monitors have a relatively

similar body form, with long necks, well-developed limbs, strong claws, and powerful tails. Varanids are a strictly Old World group, occurring throughout Australia, Asia, and Africa. By far the majority of species are Australian, and this region is regarded as the main center of monitor evolution. Like beaded lizards, the monitors are strictly oviparous (egg-laying).

The largest living lizard species, the Komodo monitor (or Komodo dragon) *Varanus komodoensis,* is a member of this group. It may grow to about 3 meters (10 feet) in total length and weigh as much as 165 kilograms (364 pounds). The Komodo dragon, which was not described scientifically until 1912, occurs only on Komodo and neighboring islands in the Lesser Sunda Chain of Indonesia, where it may be locally abundant. Like all monitors, it is active during the day. Juveniles take a variety of relatively small prey items, while adults feed on carrion and fresh prey. Eggs, lizards, and small mammals are eaten, but larger mammals, especially deer and even water buffalo weighing 500 kilograms (1,100 pounds) are very important sources of food for larger individuals. The large recurved teeth of the

▲ *The largest of all lizards, the Komodo dragon is restricted to Komodo and a few neighboring islands in central Indonesia. Although they feed largely on carrion, these huge monitors also prey on animals as large as deer and water buffalo, and there are several well-documented cases of fatal attacks on humans.*

▶ *About 25 of the 34 species of monitors occur in Australia, where they are generally known as goannas. Most inhabit arid or semi-arid country, are active by day, and are mainly terrestrial, although they are accomplished tree-climbers. This is* Varanus panoptes, *a little-known species of the north and west of Australia.*

David Curl

dragon are serrated and effectively slice through prey tissues. Powerful forelimbs and claws, used chiefly in digging, may also be used to disembowel large mammals. Attacks on humans are rare but do occur and sometimes result in death.

Other *Varanus* species are also carnivorous, taking mammalian or reptilian prey, but most

varanids are largely insectivorous. Gray's monitor *V. olivaceus* of the Philippines, however, is a specialist, subsisting on fruit and mollusks.

Even larger monitors lived in the Pleistocene (2 million to 10,000 years ago) of Australia. *Megalania prisca* was a giant goanna that reached 7 meters (23 feet) in length and may have preyed upon some of the giant marsupials that are now extinct. The largest lizards ever to have lived were also relatives of the monitors, but these were marine giants known as mosasaurs.

Not all varanids are large, however. Some Australian species, such as the short-tailed monitor *V. brevicauda*, reach only 12 centimeters (4¾ inches) head–body length. Most monitors are terrestrial, but the emerald tree monitor *V. prasinus*, a bright green species inhabiting New Guinea and the northern tip of Australia, is an arboreal specialist with a long prehensile tail. Another Australian species, Mertens' water monitor *V. mertensi*, is one of several highly aquatic species. In such water-loving species the tail is greatly compressed and is used to generate powerful lateral undulations in swimming. It may also be used to corral fish or other aquatic prey.

During the breeding season male monitors engage in combat for access to females. In actions reminiscent of those seen in some snakes, they raise up on their hind legs and tail base, and "wrestle". The contest is over when the victor has pushed over his opponent. Monitors rely heavily on chemical senses and frequently tongue-flick to test for airborne cues; the tongue is slender and deeply forked, closely resembling that of snakes. Although sight and hearing are also keen in monitors, courtship and mating behavior is primarily based on chemical and tactile cues. Varanids lay 7 to 37 eggs, typically in the soil or in tree stumps or hollows. The Nile monitor *V.*

niloticus, which ranges throughout most of Africa, may lay its eggs inside termite mounds; the architecture of the mound provides ideal humidity and ventilation, while the hardness of the mound surface protects the eggs from foraging predators. Although the adult does not guard the eggs, it may return to release the hatchlings from the termite mound. Similar behavior is known to occur in a number of arid-zone monitors.

Knob-scaled lizards

The knob-scaled lizards (family Xenosauridae) include two genera, widely separated geographically and perhaps not particularly closely related to one another. Lizards in both genera have bumpy scales and bony osteoderms in the dermis of the skin. All five species are moderate in size, ranging from 10 to 15 centimeters (4 to 6 inches) head–body length.

Xenosaurus is a genus of four species distributed from Mexico to Guatemala. They occur in a variety of habitats from semi-arid scrub to high-elevation cloud-forest, although they prefer moist or even wet habitats. All are believed to be nocturnal or crepuscular (dawn- and twilight-active), terrestrial, and secretive, seeking out holes and crevices. The common Mexican knob-scaled lizard *Xenosaurus grandis*, the best known species, occupies crevices in rocky outcrops or cavities in trees, often in forested regions; it feeds on termites and other insects, and the female gives birth to two to six live young. Xenosaurs have a

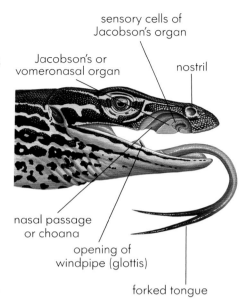

sensory cells of Jacobson's organ

Jacobson's or vomeronasal organ

nostril

nasal passage or choana

opening of windpipe (glottis)

forked tongue

▲ In snakes and some lizards the sense of smell is greatly enhanced by the Jacobson's or vomeronasal organ, located on the roof of the mouth. Although present in most reptiles, this organ is particularly well developed in species with bifid, or forked tongues. With the mouth held closed, the tongue is extended and flicked repeatedly, picking up chemical scents which are then detected by the highly sensitive sensory cells lining the Jacobson's organ.

◄ The largest of the Australian goannas is the perentie Varanus giganteus. The specimen illustrated has inflated its neck as part of its threat display. Highly agitated varanids will also arch their tail and, unlike crocodilians which are falsely reputed to do so, will often wield it as a weapon.

159

Anthony Bannister/NHPA

▲ *The armadillo girdle-tailed lizard relies on its heavily armored tail for protection, either when wedged in rock crevices or, when threatened in the open, by clamping it in its mouth to shield its more vulnerable underparts.*

gestation lasts seven or eight months, and as many as eight young may be born.

GROUP TWO: THE SCINCOMORPHS

Relationships among the scincomorph lineage have not been fully studied. The families included are the cordylids, gerrhosaurids, dibamids (tentatively), lacertids, skinks, teiids, microteiids, and xantusiids. They account for nearly half the species of all lizards. Scincomorphs have pleurodont dentition—unsocketed teeth on the inner face of the jaw bones—and rarely show crests or other ornamentation. Limbless forms are especially numerous in the group. Both visual and chemical signals are used during interactions between individuals of the same species and other species.

Girdle-tailed lizards

The family Cordylidae contains 45 species, all found in Africa. In the Eocene epoch, 57 to 37 million years ago, lizards in this family also occurred in Europe. All species are heavily armored, with bony osteoderms (bony plates), and all are day-active sun-lovers. The scales are usually large and rectangular and are arranged in regular rows around the body.

Girdle-tailed lizards are mainly rock-dwellers and live in semi-arid to highly arid areas. Their bodies and heads are flattened, and most are fully limbed. Species in the genus *Cordylus* typically have whorls of spines on the tail, which aid in active defense and in wedging the animal into rock crevices. Most species are drab, with browns, blacks, and straw colors dominating. These lizards are live-bearing, and although only one to six young are born in a season, the chance of survival is generally good. *Cordylus* species are late to reach maturity and may live for more than ten years (25 years in captivity). The armadillo girdle-tailed lizard *C. cataphractus* is a large, insect-eating species that responds to attack by curling into a ball and grasping its own tail in its mouth, thereby forming an armored ring in which only the osteoderm-reinforced dorsal surface is visible to the potential predator. Groups of related individuals often live together in deep rock cracks in the South African arid zone. Another large member of the genus, the sungazer or giant cordylid, *C. giganteus,* which reaches 40 centimeters (15¾ inches) in total length, is highly unusual because it lives in burrows rather than rock crevices and may form large colonies in the highveld savanna of the Free State and adjacent areas of South Africa. The burrows are as much as 2 meters (6½ feet) long and give protection from predators and the weather. Like other girdle-tailed lizards in temperate areas with cold winters, sungazers become inactive for part of the year.

One or two large young (up to 15 centimeters, or 6 inches, total length) are born at the end of

somewhat flattened body form in association with their rock-dwelling habits.

The crocodile lizard *Shinisaurus crocodilurus* is a semi-aquatic lizard known only from a series of isolated localities in Guanxi province, southern China. It has a rather high, laterally compressed head. Its name comes from the large tubercles on its back, arranged in rows and extending onto the tail, giving the animal a crocodile-like appearance. Until recently almost nothing was known about this species, and it was regarded as one of the rarest lizards in the world. Its food consists of fish, tadpoles, and insect larvae captured in flowing water, although terrestrial prey may be an important dietary component too. The crocodile lizard is active during the day or at twilight and spends the night perched on branches overhanging streams. If threatened it may retreat to the water. Like the Mexican knob-scaled lizards, the crocodile lizard is viviparous;

summer in February–March. Because it inhabits land capable of supporting agriculture, the sungazer has had to compete with humans for space. It is threatened but is legally protected.

Not as spiny as the girdle-tailed lizards, but better adapted for life in narrow rock crevices, are the flat lizards (genus *Platysaurus*). These are covered with small granular scales and are fully limbed. The entire body and particularly the head is amazingly depressed. Flat lizards occupy crevices and spaces beneath rock flakes on isolated rock outcrops or boulders. They are especially diverse in Zimbabwe and northeastern South Africa. Large numbers of individuals are frequently found together in retreats. In many species there is a striking sexual dichromatism, with males displaying vibrant colors (usually reds and blues or greens) that are species-specific. Females and juveniles of all species lack bright colors but have a striped back. Flat lizards are the only egg-laying members of the family Cordylidae. Two elongate eggs are laid in the moister microclimate of deep rock cracks, and a single site may be shared by ten or more females.

The grass lizards (genus *Chamaesaura*) have only tiny, spike-like hind limbs, used for stability when at rest. They have a very long tail and are "grass-swimmers", moving by pushing the body against vegetation in the high grassland habitats, and their extreme length allows for their weight to be distributed across many individual plants.

Plated lizards
Closely related to cordylids are the gerrhosaurids, or plated lizards. There are about 30 species in

this group, divided into four African and two Madagascan genera. Their bodies are generally not spiny, as in the girdle-tailed lizards, and they have a prominent lateral fold similar to some of the anguid lizards. All of them are probably oviparous (egg-laying), although little

▲ *Living on rocky outcrops in arid regions of southern Africa, the flat lizards have an extraordinarily flattened shape to facilitate hiding in crevices. The male dwarf flat lizard* Platysaurus guttatus *illustrated is in breeding colors. Surprisingly, the underside is the most colorful feature of these lizards, as males display to each other by elevating their forequarters.*

Anthony Bannister/NHPA

◄ *Early Afrikaans folklore held that the sungazer lizard faces always into the sun. Terrestrial and day-active, members of this group (genus* Cordylus) *are sometimes known as zonures.*

is known about many species. The giant plated lizard *Gerrhosaurus validus* is the largest species in the group, reaching a total length of 70 centimeters (27½ inches). It occupies rock cracks in boulder piles of the African savanna, and feeds not only on insects and other arthropods, but also small vertebrates and plant material, especially fruit. Two to five eggs are laid in midsummer, and the large hatchlings emerge about 11 weeks later. The desert plated lizard *Angolosaurus skoogii* differs from all other plated lizards in its very specialized habitat preferences. It is an inhabitant of the sand dunes of the northern Namib Desert, where it lives in small colonies and forages on the slipfaces of the dunes for insects, plant debris, and small desert melons. The body is cylindrical and the snout is wedge-shaped, with the rim of the lower jaw hidden when the mouth is closed. The toes bear broad rectangular fringes. When disturbed the lizard dives into the dune and disappears from sight.

Species in the genus *Tetradactylus* have a range of body forms, from fully limbed to nearly limbless (that is, those with small hind limbs only). Unlike certain other reduced-limbed lizards, however, these gerrhosaurs all retain large fully-functional eyes, indicating that they are surface-active rather than burrowing. Like the cordylid grass lizards described above, they have long tails and appear to "swim" across the grass of their habitats. One species, *T. eastwoodae*, has apparently become extinct because its habitat has been destroyed by humans.

The Madagascan gerrhosaurs (*Zonosaurus* and *Trachyleptychus*) are poorly known, but some species may live near water. They are somewhat similar to the mainland African plated-lizards in appearance and probably in ecology.

Dibamids

The 11 species in the family Dibamidae are among the least well-known lizards. At various times they have been regarded as closely related to anguid lizards, geckos and pygopodids, skinks, amphisbaenians, and snakes. Dibamids are small (less than 25 centimeters, or 10 inches, total length), long-bodied burrowers, with highly reduced limbs. Females lack limbs all together, while males have tiny flap-like hind limbs. The scales of the body are overlapping, and those of the snout and mandibles are enlarged and plate-like. The eyes are small and covered by a scale, and there are no external ears. The genus *Dibamus*, which is distributed through Southeast Asia, the Philippines, and the islands of the Indo-Australian archipelago (including the western part of New Guinea), accounts for all but one of the species in the family. Most species have been found in the soil or under debris in humid forests, but at least one from Vietnam has been collected in a tree! *Anelytropsis papillosus*, a similar animal, lives in rather more arid habitats in northeastern Mexico. Dibamids are insectivorous and oviparous. The eggs have a calcified shell, a feature that apparently evolved for similar ecological reasons in the Gekkota.

Lacertids

The family Lacertidae, sometimes referred to as the "true lizards", are an Old World group. Of more than 215 species, most inhabit the Mediterranean region, although others extend southward to the tip of Africa, eastward to Japan,

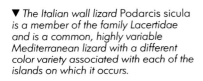

▼ The Italian wall lizard *Podarcis sicula* is a member of the family Lacertidae and is a common, highly variable Mediterranean lizard with a different color variety associated with each of the islands on which it occurs.

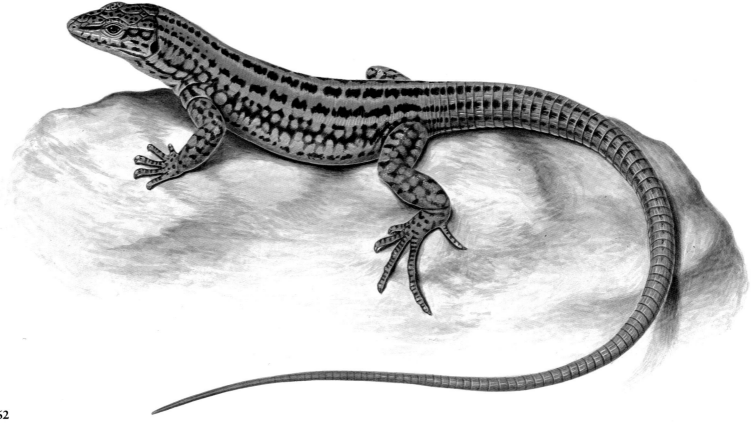

LIZARDS WITHOUT LIMBS

The reduction or loss of limbs and the elongation of the body are common and recurrent trends throughout lizard evolution. Of the 25 families of lizards, six include species that have greatly reduced limbs and two (the Dibamidae and Pygopodidae) include only reduced-limbed animals. In the skinks alone, loss of skeletal elements in the limbs has occurred about 30 times.

The origin of limblessness may be associated with two different types of habitats. In the case of dibamids, teiids, skinks, and some pygopodids, limb loss appears to correlate with the occupation of underground habitats. Anguids, cordylids, and some other pygopodids are surface-active, and their reduced limbs may have evolved during their association with dense grassland vegetation or other low thick plants, or similarly complex landscapes.

Limb loss almost certainly evolves by stages. Skeletal elements in the limbs may be lost gradually—initially only a single toe or finger may be reduced. Or it may occur more drastically, with the loss of numerous elements through the disturbance of developmental pathways. The same environmental demands that give advantages to smaller limbs also favor a more elongate, snake-like body, and in general the greater the reduction of limb elements, the greater the number of vertebrae present.

Although these "stages" can only be inferred, there are several genera that exhibit a variety of limb reduction in different species; *Bachia* among the tropical American microteiids, *Tetradactylus* among the cordylids of Africa, and the Australian *Lerista* among the skinks are three of the best examples. The genus *Lerista* includes more than 60 species and gives a very complete view of the stages in the evolution of limb reduction. In nearly all reduced-limbed lizards it is the forelimbs that disappear first. Only in certain species of the microteiid *Bachia* are the forelimbs retained and hindlimbs lost.

Even in those species that have no limbs at all, at least some remnant of the limb girdles

Mark Newton/AUSCAPE International

▲ Burton's legless lizard Lialis burtonis occurs in a range of habitats virtually throughout Australia and New Guinea. Like all pygopodids it lays a clutch of two eggs.

J.A.L. Cooke/Oxford Scientific Films

◄ Pygopus lepidopodus *is one of the largest and commonest of Australian "legless" lizards. This close-up shows its rudimentary hind limbs.*

remains internally. Pygopodids, for example, retain vestiges of the pelvic girdle, although the connection of the girdle to the vertebral column is lost. Remnants of the pelvis and femur are retained even in some snakes.

and north to the Arctic Circle. With a few exceptions the family is reasonably uniform in body form and structure. Head scales are large and often contain bony osteoderms; body scales are usually small and granular on the back but large and rectangular on the underside. Most species are terrestrial, and all but a few are exclusively diurnal—lacertids prefer high body temperatures, and many are inactive when weather conditions are unfavorable.

Wall lizards (genus *Podarcis*) are familiar to most Europeans, as they are conspicuous in both

natural habitats and those influenced by humans. The common wall lizard *P. muralis* is the most widely distributed species in the genus. It is typical of many lacertids in preferring somewhat open country, where it may be found climbing on rocks, trees, or buildings. Typical of the family, males are generally patterned differently from the females. In the more ground-living lizards of the genus *Lacerta*, males are especially distinctively colored, often featuring greens in contrast to the browns of females. Many species of lacertids display highly variable color and/or scalation in different locations, especially island populations of some species. Five species of *Lacerta* consist of females only, which reproduce parthenogenetically (without a male). The species common in Europe are mainly insect-eaters, although larger species such as the ocellated lizard *L. lepida* also eat vertebrates and fruit.

The most widespread lacertid, and perhaps the most widespread of all lizards, is the viviparous lizard *L. vivipara*—sometimes placed in its own genus, *Zootoca*—which grows to only 65 millimeters (2½ inches) head–body length. As its name suggests, this lizard gives birth to live young, a trait shared with only two other lizards in the family. This feature has allowed the lizard to succeed in harsh climates that could not be colonized by egg-laying species. By retaining the developing embryos within the body, the female is able to regulate the temperature and moisture more closely than if the embryos were left to hatch in some protected site. Interestingly, in the warmer south of its range in Spain and adjacent France, the viviparous lizard is reputed to lay eggs. The occurrence of two reproductive modes in one species highlights the relative ease with which viviparity may evolve from egg retention. Litter size in the viviparous lizard is generally four to eleven. Clutch size in other lacertids is roughly related to size, with smaller species producing one or a few eggs, and larger forms laying ten or more. Mating occurs in spring, and eggs are laid by early summer, hatching in time for young to feed before winter hibernation or inactivity.

The largest living lacertid, the Canary Island lacertid *Gallotia stehlini*, at 46 centimeters (18 inches) is considerably smaller than its now-extinct relatives, also on the Canary Islands. These giants reached sizes of more than 1 meter (39½ inches) total length. Apart from their large size, *Gallotia* species are the most vocal of lacertids and one of the few groups active during twilight or at night.

Members of the genus *Takydromus* range through eastern Asia. They are fully limbed, but like *Tetradactylus* and *Chamaesaura* in the families Gerrhosauridae and Cordylidae they have extremely long tails and are able to "swim" through dense groundcover.

Very few lacertids have adapted to life in tropical forests, but several major groups of lacertids have become specialized for a sand-dwelling existence. The most extreme case is that of the shovel-snouted lizard *Meroles anchietae*, endemic to the coastal Namib Desert. Like many dune lizards, its wedge-shaped head features a countersunk lower jaw, and its feet are strongly fringed. This species is an evolutionary extreme in a genus that, in general, shows some degree of modification for sand-dwelling. *Acanthodactylus*, a genus from North Africa and the Middle East, shows many of the same characteristics and occupies comparable habitats in the Sahara and its surroundings, although these lizards are often

▶ *The viviparous lizard* Lacerta (Zootoca) vivipara *forms eggs, but these are generally retained in the body until the time of hatching, resulting in the birth of miniature replicas of itself, a habit otherwise unusual in the lacertid lizards. This hardy and widespread species occurs in Europe and northern Asia.*

L. Campbell/NHPA

◄ *With its spade-shaped snout and underslung jaw, the sand-diving lizard Meroles cunierostris is one of a number of closely related lacertid lizards adapted to loose sand conditions in the deserts of southern Africa.*

Anthony Bannister/NHPA

more generalized in their ability to use a variety of microhabitats. Another African species, the bushveld lizard *Heliobolus lugubris*, is largely a termite-eater in grassy savannas and open country. Although the adults are not especially strikingly patterned, the juveniles are black with conspicuous white markings. They may be seen walking in a scuttling manner with the back arched high. This odd behavior is a defensive mechanism, as the juvenile in this posture resembles the oogpister, a noxious beetle occurring in the same habitat.

Skinks

The family Scincidae contains more than 1,300 species in about 100 genera, making it the largest of all lizard families. Skinks are cosmopolitan in their distribution, although there are relatively few in Europe, northern Asia, South America, or much of North America. Many skinks are secretive and inhabit leaf-litter, rock crevices, or rotting logs. Limb reduction, a common trend in many lizard

families, has occurred in numerous groups of skinks and a variety of limbed and limbless forms are burrowers. Nearly all skinks are covered by relatively smooth, overlapping scales, giving them a fish-like appearance, as reflected in such common names as "sandfish" (*Scincus*) and "land mullet" (*Egernia*). Most skinks are primarily gray or brown in color, although bright colors are sometimes seen in males and juveniles. Many of the more secretive forms are nocturnal or crepuscular (twilight-active), but many others are diurnal (day-active).

Four subfamilies of skinks are recognized: feylinines, acontines, scincines, and lygosomines. Feylinines are a small group of limbless burrowing skinks found only in tropical central and west Africa. The six species are apparently all viviparous (giving birth to live young).

The acontine skinks, about 20 species of limbless African lizards, are more widespread and occupy a wider range of habitats. Many occur in arid regions of southern Africa, where they live in

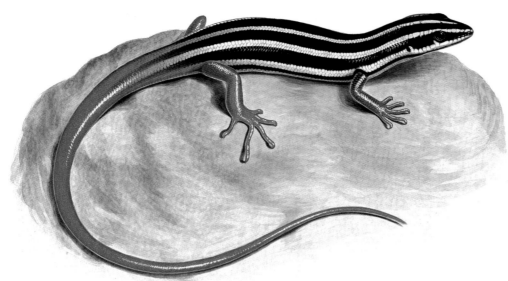

young of the five-lined skink E. *fasciatus* have bright blue tails, although the coloration is lost as the animals grow. Tail loss is common among skinks, and in many populations most adults have regenerated this appendage. These skinks also show sexual dimorphism: males are larger than females and lose the patterning seen in juveniles, whereas this is retained by females. In the breeding season males of the five-lined skink develop orange color on the head, and the width of the head increases as the cells of the jaw muscles increase in size (hypertrophy). Males recognize each other visually and chemically and engage in combat for territories and access to females. There is little courtship display in most skinks, and tactile and chemical information is exchanged before copulation in most cases.

Many of the remaining genera of scincine skinks are small groups living in Africa, Madagascar, the Philippines, and the Middle East.

▲▶ *Like many other lizards, most skinks can discard their tail as a defense mechanism. As a consequence, many species have evolved brilliantly colored tails to deliberately draw the predator's attention away from the vulnerable head and trunk. This adaptation is often accompanied by a striped body pattern, a confusing target for a predator. Compare the unrelated red-tailed skink* Morethia ruficauda exquisita *(above), an Australian lygosomine, and a juvenile five-lined skink* Eumeces fasciatus *(right), a scincine from North America.*

sandy soils. The Cape legless skink *Acontias meleagris* of South Africa may be found beneath rocks or surface debris; if exposed it rapidly dives downward into the soil. In addition to having lost their limbs, these skinks have large headshields that are used in burrowing and, as is typical of many burrowing skinks in all subfamilies, the eyes are small and there is no external ear or exposed tympanum. The giant legless skink *A. plumbeus* may reach a length of 55 centimeters (21½ inches) and gives birth to as many as 14 young. Smaller viviparous species usually have only one to four young in a litter.

The scincines may be the most primitive living skinks. They are widely distributed, except for Australia and the Pacific islands, and members of the genus *Eumeces* are the most numerous skinks in North America. Some, such as the Great Plains skink *E. obsoletus*, show a high degree of parental care, and females remain with the eggs during the incubation period; they may protect the eggs from predation and/or from mold or spoilage. The

Among the reduced-limb forms are the almost 20 species of the African genus *Scelotes*, which show a variety of stages in limb loss. The sandfish (genus *Scincus*) are also members of this group. These are sand-swimming lizards of North Africa and southwest Asia. Their modifications for life within the sand, where they spend most of their time, include a countersunk lower jaw, exceptionally smooth scales, and protected ears. Unlike *Scelotes*, the sandfish are egg-layers.

The lygosomines are the largest group of skinks and account for all of the species occurring in Australia and islands of the Pacific Ocean, as well as many Asian and African and a few Central and South American species. Members of the *Mabuya* genus occur on several continents, but it is probable that not all species are closely related. Most species are placed in the genus because they lack the more distinctive features characteristic of other groups; therefore, most are rather generalized skinks, either terrestrial or rock-dwelling and usually insect-eating. Many of the

African skinks are classified in this genus. They include a large number of largely rock-dwelling forms, as well as some specialized species. The wedge-snouted skink *Mabuya acutilabris*, for example, occurs in southern Africa where it occupies burrows at the bases of grass tussocks and bushes in areas of sandy soil. It is more active than many other skinks and behaves somewhat like the lacertid lizards, which occur in the same area. Reproductive mode varies in the genus, and in the Cape skink *M. capensis* both live birth and

egg-laying have been documented. In the South American Brazilian skink *M. heathi*, the fertilized egg is very small at first, and the developing embryo receives nutrition not from yolk but from the mother via a mammalian-like placenta. This is the most highly developed example of live birth in all reptiles.

Among the lygosomines of the Australian region are a group of moderate to very large skinks, allied to land mullets (genus *Egernia*) which are among the most bizarre lizards in the family. One of the

▲ Feeding mainly on worms, the silver sand lizard or two-legged skink Scelotes bipes *is common across much of southern Africa.*

▼ A relative of the blue-tongued skinks, the Australian pink-tongued skink Cyclodomorphus gerrardii *has a blue tongue as a juvenile, but adults display a bright pink tongue when threatened.*

Jean-Paul Ferrero/AUSCAPE International

▲ *So-called from its spiny scales, the crocodile skink* Tribolonotus gracilis *inhabits New Guinea.*

▼ *Big and burly, the eastern blue-tongued skink of Australia is likely, when threatened, to stand its ground, puff itself up, hiss, and stick out its brightly colored tongue.*

C.B. & D.W. Frith/Bruce Coleman Limited

strangest of these is the Solomon Islands prehensile-tailed skink *Corucia zebrata*. This large tree-dwelling lizard feeds on plants at night and gives birth to a single large young. Related to the prehensile-tailed skinks is the Australian blue-tongued skink *Tiliqua scincoides*, one of the largest skinks, reaching 33 centimeters (13 inches) head–body length. Like other very large skinks it is partially herbivorous, eating plant material as part of a broad diet. Live birth has evolved

independently in at least 22 lineages of skinks, most frequently in lygosomines. In the blue-tongued skink about ten young are born, whereas in the shingleback *T. rugosa* only two very large twins comprise the litter. The shingleback is also huge in comparison to most skinks and it is distinguished by the large rough scales that cover its body. In this species the tail is especially short and resembles the head; it may serve to disorient predators and misdirect attacks. Another genus in this group is *Tribolonotus*, the spiny skinks of New Guinea and the Solomon Islands. These are the least "skink-like" members of the family, in that the scales are uniquely covered with bumpy tubercles.

Another group of lygosomines, including *Emoia* (72 species) and *Cryptoblepharus*, is also chiefly distributed in the Australian region. These are among the only lizards other than geckos to have successfully colonized the islands of the central and eastern Pacific. Also included in this group are the genus *Carlia* (21 species in Australia and more elsewhere) and all of the skinks of New Zealand (more than 30 species).

The third group of lygosomines includes many sand-swimmers or burrowing forms from Australia, New Guinea, and Asia. Among these are *Ctenotus* (about 80 species); and *Lerista*, a genus of 60 Australian species of skinks with small or reduced limbs, mostly sand-swimmers inhabiting the arid regions of the continent. Although the small limbs (if present) may be used when on the surface, movement in the sandy soil is by lateral (side-to-side) undulations of the body. *L. bougainvillii* is one of the most fully limbed forms, while *L. apoda* has lost all external traces of both fore- and hind limbs.

Tropical Asia is home to a number of tree-dwelling lygosomine skinks. One of these, the green-blooded skink *Prasinohaema virens*, which also occurs in the New Guinea region, is particularly noteworthy because it has scansors on its toes (a development parallel to geckos) and green pigment in the blood. The mucous

AUTOTOMY: SHEDDING THE TAIL

Most lizards are relatively small and cannot repel attackers by using force, although many will bite, lash with the tail, cry out, or defecate when attacked. Another mechanism of escape is autotomy, the loss of body parts. Although autotomy is common in many invertebrates, it occurs in vertebrates only in some salamanders, a few mammals (several rodents), and the majority of lizards. At least some lizards of most families except the Agamidae, Chamaeleonidae, Helodermatidae, Lanthanotidae, Xenosauridae, and Varanidae, may drop the tail or part of it when grasped by a predator. In many cases the tail continues to wriggle long after it is detached, perhaps serving to distract the predator and thus buying time for the lizard to escape.

Tail loss may carry a high cost. The tail is a common site of fat deposition in lizards, and loss of these important stores may decrease survival rates during winter and/or reproductive output. In addition, the lizards must cope with the temporary loss of any specialized tail functions, such as those in locomotion, grasping, or social behavior. Males of the side-blotched lizard *Uta stansburiana* (a phrynosomatid) that are deprived of their tails may even suffer a decrease in social status and thus lose the opportunity to breed.

In most lizards capable of autotomy, the rupture of the tail occurs not between adjacent vertebrae but within a single vertebral unit. A zone of weakness, developed prior to birth or hatching, runs through each of the autotomic vertebrae. This division corresponds to a boundary between two muscle segments and is continued to the surface, where in this region the skin is somewhat weaker. Blood loss is minimal in autotomy, and a lizard that escapes will regrow the tail over a period of months, although the lost vertebrae are replaced by a cartilaginous rod and the muscles and scales that regrow are generally irregular. The dwarf day geckos of Africa (genus *Lygodactylus*) and some other species have specialized tail forms, such as adhesive tail pads similar to those on the toes. Even these, with only minor modifications, are faithfully reproduced by the regeneration process.

Autotomy of a different type characterizes some geckos. In the bronze gecko *Ailuronyx seychellensis* of the Seychelles and the Madagascan genus *Geckolepis*, as well as some others, the skin on the animal's back is weakened by gaps in its fibrous structure. When predators grasp the lizard, the skin (or rather, most of the skin, for a thin layer

Michael and Patricia Fogden

remains to protect the underlying tissues) rips away. As in tail autotomy the loss of part of the body may distract a predator long enough to allow the lizard to escape.

▲ *A tokay gecko Gekko gecko with a regenerated tail.*

membranes, bones, and eggs of this skink are also green. The pigment is similar to that found in bile (a digestive fluid) which is also green. The function of this pigment, if any, is unknown.

Macroteiids

Macroteiid lizards (family Teiidae) are remarkably similar in appearance to the lacertids, and they are geographically complementary—lacertids occur in the Old World, and macroteiids in the New World (the Americas). There are about 105 species in nine genera. All macroteiids have well-developed limbs and are primarily terrestrial and day-active. Head scales are usually large, dorsal scales are tiny, and ventral scales are enlarged and rectangular. Egg-laying is the universal mode of reproduction in the group.

Whiptail lizards (genus *Cnemidophorus*) are the only macroteiids in the United States and are the smallest members of the family. Whiptails are active foragers, at times moving almost constantly in search of prey. Insects and other arthropods are the principle components of the diet. Prey such as termites are often excavated by digging. The western whiptail *C. tigris*, from western North America, is typical in its preference for high temperatures. This lizard is active at body temperatures of about 40°C (104°F). As a result its metabolic rate is high, as is the demand for more energy (which means more food). During periods of inclement weather and the colder months of the year the lizard retreats to a burrow, usually near the base of a shrub or bush. The female lays small clutches of eggs, but may produce two clutches per year. Low clutch size (one to six eggs) is the rule for whiptails as a whole. While

▼ *The checkered whiptail* Cnemidophorus tesselatus *(below) is a large teiid native to North America. Like many whiptails, this is an all-female species, able to lay fertile eggs in the absence of males. Essentially a large version of the whiptails, the jungle runner* Ameiva ameiva *(bottom) is a wideranging species consisting of several races spread throughout much of Latin America.*

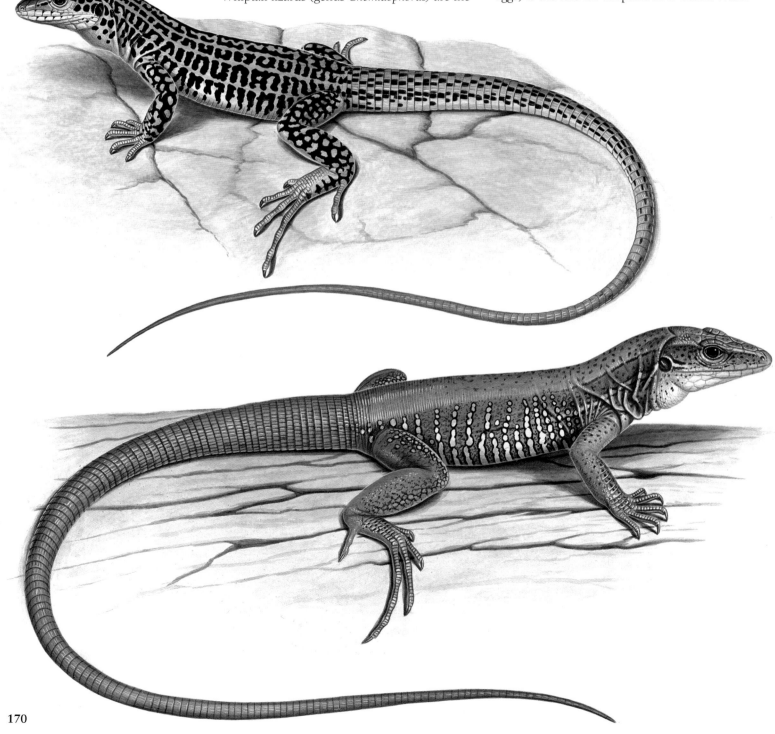

the western whiptail is a bisexual species, parthenogenesis (reproduction without fertilization) characterizes at least 20 species, some as yet undescribed. *Cnemidophorus* species are widely distributed in North America and extend south to Argentina. They are joined in Mexico and further south by the genus *Ameiva,* the jungle-runners. Both genera occur in a variety of habitats from desert to tropical forests. Some species in both groups are sexually dichromatic, with the males having brightly colored markings on the back and sides, used in displays to rival males and in courtship.

Larger macroteiids have a broad diet and will eat insects, vertebrates, and even some plant material. *Callopistes* inhabits valleys in the Andean region and along the coast of Chile and Argentina and is a specialist feeder on other lizards, mostly iguanids. One of the largest species, the tegu *Tupinambis teguixin,* occurs throughout much of tropical and subtropical South America and is primarily a predator on small mammals. Some tegus grow almost as big as the more massive monitor lizards, with head–body lengths of up to 45 centimeters (17¾ inches). As many as 32 eggs, a record for the family, may be laid by females of this species. The caiman lizard *Dracaena guianensis* rivals the tegu as the largest macroteiid, with a total length of up to 1.3 meters (4¼ feet). It has unusally large, flattened teeth, which it uses to crush the shells of mollusks, its primary food. It is one of a few semi-aquatic macroteiids and has a compressed tail and large tubercles on the back.

Many of the larger macroteiids are captured for food by rural people in South America. Tegus are especially valued as meat and for their skins, which are sold commercially to be made into fashionable lizard-skin leathergoods.

Microteiids

Microteiid lizards (family Gymnophthalmidae) are far more structurally diverse than macroteiids, with which they are grouped by some scientists. Like macroteiids, all are oviparous (egg-laying) and there are some parthenogenetic (unisexual) forms. There are 140 species in 30 genera, ranging from Mexico southwards through South America, achieving their greatest diversity in tropical South America. Most are small, leaf-litter-dwelling species that are active at night or intermittently active in their sheltered microhabitat throughout much of the day. Like many other groups, the microteiids include reduced-limb forms; these occur in eight different genera, suggesting that reduction has been a recurrent theme in microteiid evolution. Members of the genus *Bachia* show a wide range of stages in reduction. Unlike other lizards (but in common with the amphisbaenian *Bipes),* the hind limbs are reduced more than the forelimbs. *Bachia trinasale* possesses only two small forelegs, the hind limb rudiments being entirely internal. Members of this

Jany Sauvanet/NHPA

genus are burrowers, digging tunnels in the forest soil beneath leaf-litter. If threatened on the surface, however, they may leap erratically. In contrast, the litter-dwelling microteiids of the genus *Echinosaura* "freeze" when threatened and rely on camouflage to avoid predation. Lizards in genus *Neusticurus* are water-loving and prefer swampy areas within the tropical forest. When disturbed they retreat to water. Most species of microteiids prefer shaded conditions and are most active in the rainforest or on overcast days. *Cercosaura ocellata,* however, seeks out sunny spots on the forest floor, as do the very skink-like genera *Tretioscincus* and *Gymnophthalmus.*

▲ *The common tegu* Tupinambis teguixin *is one of the largest members of the family Teiidae, a group of energetic, terrestrial lizards confined to the Americas.*

Night lizards

The night lizards (family Xantusiidae) are a group of 19 species of New World lizards. The name may be somewhat misleading, since it has been found that some species are active by day, but usually in hidden microhabitats. The species that have been studied generally have a low preferred body temperature, averaging 23°C (73°F), nearly 10 degrees lower than most lizards. All species are viviparous, giving birth to one to eight live young after a gestation of three months. Night lizards have been considered as relatives of geckos, with which they share certain features, such as the absence of eyelids and the presence of a protective spectacle, but they are probably most closely related to skinks, lacertids, and teiids.

The largest genus is *Lepidophyma,* with 14 species distributed from Mexico to Panama, inhabiting both tropical lowland forests and rocky semi-arid zones at high elevation. Most species have rows of enlarged tubercles on their back. Little is known about the biology of these animals, but most seem to be nocturnal or active at

▶ The tree-dwelling flying lizards (genus Draco) of the tropical Asian region have five to seven pairs of ribs, much lengthened, hinged, and connected by a membrane in such a way that when pulled forward they form a parachute. Such a surface allows glides of up to 60 meters (about 65 yards) or more, with a considerable degree of control.

Jean-Paul Ferrero/AUSCAPE International

Stephen Dalton/NHPA

▲ In the flying geckos (genus Ptychozoon) of Southeast Asia folds of skin extend along the sides of the body and tail, increasing the total surface area presented to the air and thus allowing some degree of control over the rate and direction of fall.

GLIDING THROUGH THE TREES

Several lizards have a body built for gliding. None of them can use flapping or powered flight, but they are able to slow their rate of descent and move through the air in a controlled manner. When the rate of descent is less than 45° from the horizontal this is referred to as gliding; when the angle is steeper it is parachuting. Only the flying dragons (genus *Draco*, in the family Agamidae) can glide effectively, but parachuting occurs in the "gliding" geckos (*Ptychozoon*), and the "gliding" lacertid (*Holaspis*). Each of these lizards has a way of enlarging their surface area to increase drag and thus slow the descent. In *Holaspis* the increased surface consists of fringed scales on the toes and along the sides of the tail. In *Ptychozoon* fleshy fat-containing flaps extend along the sides of the body and the tail, and webs span the spaces between the toes. In *Draco* there is an altogether different type of mechanism: five to seven ribs are lengthened and support a membrane that can be extended or folded back against the body by

muscular action. In "flight" the ribs are pulled forward, stretching out the membrane to provide the maximum surface area for lift.

Some lizards, such as the green anole *Anolis carolinensis* and the butterfly agamid *Leiolepis belliana*, have no structural modifications for parachuting but are behaviorally equipped to assume a posture that maximizes air resistance when they fall (or are dropped) from a significant height.

Gliding or parachuting may be a mode of escape or a means of locomotion. The distance covered by a gliding gecko when it parachutes is relatively short (less than 10 meters, or 33 feet), but it allows the animal to rapidly leave an area of potential danger or move between trees in the forest. Aerial movements of *Draco* may be more impressive, with horizontal distances of up to 60 meters (65 yards) being covered during only a small vertical descent, and fine movements of the membrane and tail give precise control of direction, speed, distance, and angle of descent.

twilight, and some occupy rotting logs or even caves. Most species are insect-eaters, but Smith's night lizard *L. smithii*, one of the cave-dwelling species from Mexico, feeds on fallen fruit from fig trees near its rocky retreats. Some southern populations of *L. flavimaculatum* consist only of females. The Cuban night lizard *Cricosaura typica* is the sole representative of a second genus of the family.

A third genus, *Xantusia*, occurs in northern Mexico and the southwestern United States. They are characterized by small granular dorsal scales contrasting with large smooth head shields and large rectangular ventral scales. These lizards have small litters but are reasonably long-lived, regularly reaching the age of five years, and as many as 12 to 15 years in the largest species. The desert night lizard *X. vigilis* was considered to be rare until it was discovered that individuals of some populations spend much of their life among fallen debris of Joshua trees and other plants, where their density may be very high. Like others in its genus, this species matures slowly and is long-lived. The granite night lizard *X. henshawi* lives in southern California, but unlike the desert night lizard, it is distributed only in regions with granite boulders, where it occupies crevices. Its litter size of one or two per year is one of the lowest for any lizard. The island night lizard *X. riversiana* occurs only on the Channel Islands off the coast of southern California, where it reaches densities of several thousands per hectare in prime habitat. The lack of competitors and the protected habitat allows enormous concentrations to build up in spite of the low reproductive rate, but the island night lizard is considered threatened because it occupies only a few small islands—the species as a whole would be susceptible to predation and habitat destruction if mammals were introduced to the islands. The island night lizard is much larger than mainland species: 90 millimeters (3½ inches) head–body length, compared to 35 to 70 millimeters (1⅓ to 2¾ inches). And unlike most other *Xantusia* species, it eats plants as well as insects.

CONSERVATION

While very few lizards are dangerous to humans, humans are most decidedly dangerous to lizards. Many of the larger lizards, especially monitors and large teiids such as the tegu, continue to be exploited for their skins, which are fashioned into leathergoods.

Accidental introduction of rats and intentional introduction of cats and other predators has greatly diminished populations of some lizards, especially on islands. In New Zealand skinks and geckos that once occupied the main islands now occur only on small offshore islands which are less disturbed. On Guam, the brown tree snake *Boiga irregularis* has caused the extirpation of both birds and lizards. Even other lizards introduced into new ecosystems may have negative effects on native forms. In several areas of the Pacific, for example, the house gecko *Hemidactylus frenatus*, introduced to many islands during and after the Second World War, seems to be causing a dramatic decline in numbers of the Indo-Pacific gecko *H. garnotii*, which had previously been numerically dominant.

Our greatest impact on lizards has certainly been the destruction of their habitats. Alteration of habitats in agriculture, recreation, and urban expansion effectively destroys most resident lizard populations, especially as most lizards cannot flee to distant undisturbed areas or even to adjacent plots of land. Only in rare cases, such as that of the Coachella Valley fringe-toed lizard *Uma inornata*, in California, have conservationists been successful in setting aside land specifically for lizard preservation. In all too many cases, lizard species are affected long before biologists understand their ecology and habitat requirements, and in some cases the lizards become extinct before there is time to scientifically describe them.

AARON M. BAUER

▼ *The granite night lizard* Xantusia henshawi *is an inhabitant of granite outcrops of southern California and Mexico's Baja California; its somewhat flattened body allows it to slide in and out of crevices easily.*

SNAKES

Many threadsnakes (family Leptotyphlopidae) and blind wormsnakes (family Anomalepididae)
Length: less than 15 cm (6 in)
Weight: less than 2 g (¹⁄₁₀ oz)

Anaconda *Eunectes murinus*
Length: up to 10 m (33 ft)
Weight: 250 kg (550 lb)

CONSERVATION WATCH
!!! The 12 critically endangered species are: Antiguan racer *Alsophis antiguae*; black racer *Alsophis ater*; St Vincent blacksnake *Chironius vincenti*; *Coluber gyorensis*; *Liophis cursor*; Kikuzato's brook snake *Opisthotropis kikuzatoi*; golden lancehead *Bothrops insularis*; Aruba Island rattlesnake *Crotalus unicolor*; Cyclades blunt-nosed viper *Macrovipera schweizeri*; *Vipera bulgardaghica*; *Vipera darevskii*; *Vipera pontica*.
!! 13 species are listed as endangered.
! 25 species are listed as vulnerable.

Many reptiles show evolutionary tendencies toward lengthening the body and reducing the size of the limbs, but it is the snakes that have developed most successfully in this way. They have diversified dramatically during recent geological history and now inhabit most parts of the planet outside the polar regions—from alpine meadows to the open ocean. They are all carnivorous but use a wide variety of methods to find and overpower their prey. Their diets are equally diverse: some species feed on the tiny eggs and larvae of ants, whereas others can eat animals as large as antelopes, tapirs, and wallabies.

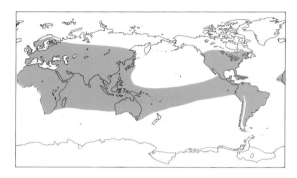

WHAT IS A SNAKE?

Many people assume that a snake is simply a reptile without legs. However, many kinds of lizards have legs so small that they are difficult or impossible to see without close inspection, making it hard to tell whether the animal is a snake or a lizard. Unfortunately there is no simple diagnostic character that is easy to use in all situations. Most lizards (although not all) have at least a vestige of the hind limbs, even if it is just a small bud or a flap of skin, whereas snakes generally do not. Most lizards have external eardrums, which snakes do not. Most lizards have relatively long tails, whereas snakes usually have short ones. Most lizards have movable eyelids, whereas snakes have a fixed transparent scale over each eye. The differences that are absolutely consistent and reliable in distinguishing between snakes and lizards are all fairly subtle, mostly involving the structure of bones in the head. This won't be of much use if you want to identify the long, thin object that you've just seen · disappearing under your house, but it does emphasize the great similarity between snakes and lizards—which is why they are placed in the same order, Squamata, within the class Reptilia.

HOW SNAKES MOVE

The most obvious distinguishing feature of snakes is their shape. Lengthening of the body and reduction or loss of limbs has occurred many times in the evolution of the vertebrates—for example, in eels, salamanders, caecilians, and lizards. This evolutionary change in body shape has profound consequences for an animal's

biology, most obviously in the way the animal moves around. Snakes have several features in their vertebral column that are related to limbless locomotion. Firstly, the number of vertebrae are greatly increased, providing a much more flexible backbone. Humans have only 32 vertebrae, whereas some snakes have more than 400. Secondly, snakes have extra projections from each vertebral element, so that adjoining vertebrae are connected more tightly, helping to provide stability to this extremely long backbone.

Even without legs, several different methods of locomotion are possible. The most familiar technique used by snakes is lateral undulation. All snakes seem to be capable of using this method when they are swimming or when they are moving over solid surfaces that have enough irregularities for them to obtain sufficient grip. The snake moves forward by pushing its body against these irregularities and can often travel quite rapidly. Speeds of up to 10 kilometers (6 miles) per hour have been reliably measured. Much faster speeds have been claimed, but none of these are anywhere near the kinds of speeds

▼ The brilliant coloration of the green tree python Chondropython viridis of New Guinea and northern Australia camouflages it among epiphytic plants in its rainforest habitat.

Jean-Paul Ferrero/AUSCAPE International

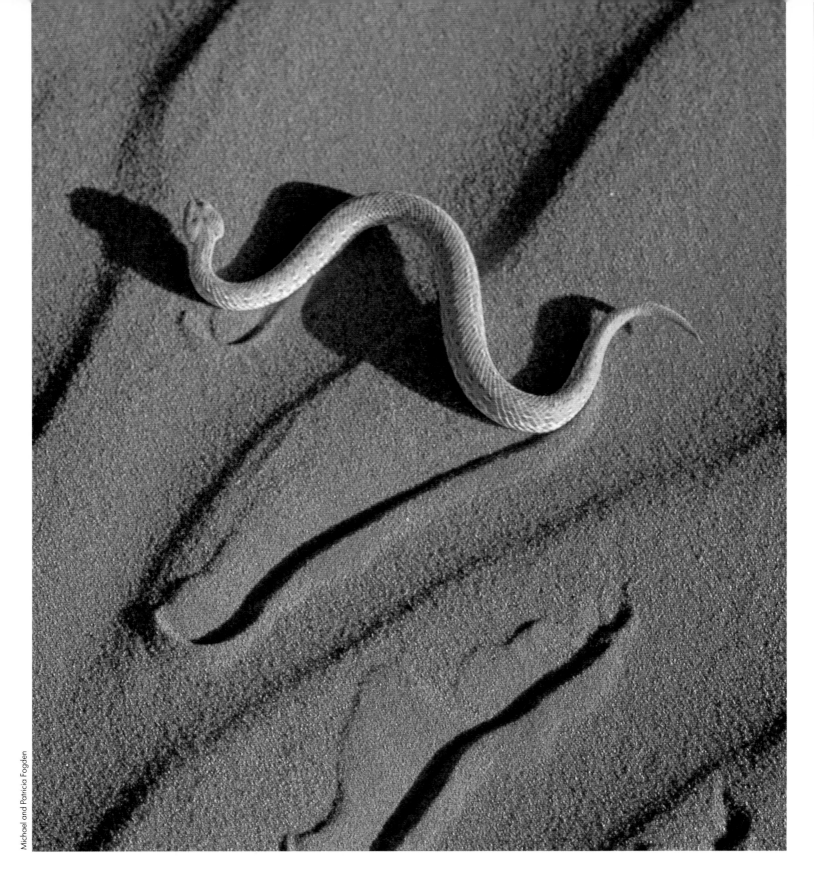

▲ *The most efficient mode of locomotion across loose sand, sidewinding has developed independently in several species of desert-dwelling vipers. Here a Peringuey's desert adder Bitis peringueyi sidewinds its way across a dune in the Namib Desert.*

often attributed to snakes in comic books. No snake in the world is capable of overtaking a galloping horse, as in one common myth. If the surface of the ground is very smooth, snakes can move by other techniques such as rectilinear locomotion, concertina locomotion, slide-pushing, and sidewinding. In rectilinear locomotion (the "caterpillar crawl") the snake stretches out in a straight line and depends on movements of the ventral skin (the underside) relative to the rest of

the body. The snake pulls itself forward by muscular contraction while anchoring its belly-scales using friction on the ground. It then pulls the ventral scales forward to a new friction point and repeats the process. This ventral anchoring and forward movement take place simultaneously at independent segments of the body. Large pythons, boas, and vipers use rectilinear locomotion frequently, especially when creeping up to prey across open ground. The snake's movements are subtle and difficult to detect when it is moving in this way.

In concertina locomotion and sidewinding, the snake uses a point of contact with the ground as purchase, then lifts its trunk clear of the ground to establish another point of contact. Sidewinding is particularly well suited to soft substrates such as sand and soft mud, where it would be difficult to find firm irregularities allowing lateral undulation. Small desert-dwelling vipers and mudflat-inhabiting colubrid snakes rely heavily on sidewinding to move around.

A few colubrid snakes of the Indo-Pacific jungles are known as "flying snakes" because of their unusual method of moving from tree to tree. They launch themselves from high trees and by flattening their bodies can glide for considerable distances without being injured when landing.

A NARROW ADVANTAGE

A snake is really just a long tube. Unfortunately, elongation means that the mouth is very small relative to the size of the body, and therefore to the amount of food required. Elongate limbless vertebrates have adopted several ways to overcome this problem. Some eat large numbers of very small prey items, which can easily be ingested even by an animal with a small head. This is the most common solution used by lizards, and a few snakes, such as wormsnakes. Other elongate vertebrates catch larger prey and tear off pieces small enough to swallow. Amphisbaenians use this approach. Most snakes, however, have a third solution: drastic modifications to the skull, to enable the snake to ingest prey items that are very large relative to its own size.

As a result the head of a snake is very different from that of other reptiles. There has been a general loosening of attachments to permit greater flexibility, so that the snake's skull contains several points at which adjacent bones can move relative to each other. Most importantly, the two halves of the lower jaw are not rigidly fused together, but instead are joined at the front by an elastic ligament that allows them to stretch far apart. The opening to the windpipe can be extruded to one side so that the snake can keep breathing while it is engaged in subduing and swallowing a large prey item, a process that may take many hours. Most elements of the snake skull are reduced, permitting greater flexibility, although the floor of the braincase is thickened and provides protection for the brain against injury from struggling prey.

This complex reorganization of skull structure allows snakes to swallow truly prodigious meals. Many snakes routinely feed on prey much larger in diameter than their own heads, and some species have been observed eating prey weighing considerably more than themselves. Popular attention has been focused on large snakes and

Michael and Patricia Fogden

▲ A spider accidentally stumbles across a blunt-headed tree snake subduing an anole.

Stan Osolinski/Oxford Scientific Films

▲ *In a spectacular demonstration of skull flexibility, an African rock python* Python sebae *distends its mouth to swallow an impala. Although many snakes habitually swallow animals larger than their heads, prey items the size of this antelope are uncommon.*

large meals, such as pythons swallowing pigs, antelopes, and wallabies, but the achievements of smaller snakes are just as impressive. Studies suggest that vipers can swallow large prey more quickly and efficiently than other snakes can.

Because they can eat such huge meals relative to their own size, snakes can survive for long periods without feeding. This ability is increased by the low rate at which snakes use energy for their own bodily processes. As ectothermic ("cold-blooded") animals, they don't need to expend large amounts of energy just to maintain a high body temperature (as do birds and mammals). Many snakes probably eat very infrequently, perhaps only a few times a year. This means that they can shrink their digestive system during times when it is not being used, which reduces their energy expenditure still further. Snakes (and other ectotherms, such as lizards) may therefore be able to survive in areas where food supply is low or erratic, which is probably one of the reasons that snakes and lizards have been more successful in many desert areas than the supposedly "advanced" birds and mammals.

Ectothermy confers other advantages as well. Because mammals and birds must maintain high stable body temperatures, they face enormous problems in losing heat to the environment. To reduce that heat loss, such animals must be

covered in insulation (fur, feathers, or blubber) and have to be fairly round in shape. Elongate bodies have a much greater relative surface area, and so a greater area over which heat can be lost to the surrounding air. Ectotherms don't face this problem, and in terms of energy loss and gain they can "afford" to be any shape at all. An elongate shape is often an advantage—for example, it allows the animal to enter narrow crevices for food or shelter—and this is probably the reason why elongate shapes are so common in ectotherms (fishes, some amphibians, reptiles) but not endotherms (birds, mammals). For the same reason—that is, the ratio of surface area to volume—warm-blooded animals must be relatively large. Ectotherms don't have to be, and so they can operate with much smaller body sizes.

The elongate shape of snakes also has advantages in terms of heat gain and loss. Because snakes, especially small ones, have such a high surface area relative to their body volume they are able to gain heat rapidly when they bask in the sun or press against a sun-warmed rock. They also have the ability to slow down the rate at which they heat or cool, either by changing shape (a tightly coiled snake has a much lower surface area through which heat can be lost or gained) or by physiological changes (especially, redirection of blood flow between surface vessels and deeper ones). The overall result is that an elongate shape gives snakes a great degree of control over their body temperatures.

FOOD AND FEEDING

Because snakes cannot bite or tear their prey to pieces (unlike most lizards), small snakes simply cannot swallow very large prey. Perhaps as a consequence of this gape-limitation, the body size of a snake has a major influence on its feeding habits. Smaller snakes eat smaller prey. In some cases this just means that juvenile and adult snakes eat the same species of prey but that the young snakes eat smaller (younger) prey individuals. This is true, for example, of many snake species with specialized diets, like some of the smaller Australian venomous snakes; they eat scincid lizards throughout their lives, starting out by eating newly hatched skinks and graduating to adult skinks as they themselves grow larger. In snakes that attain larger sizes, the increase in body size often means a change in the type as well as size of prey; for example, many pythons and vipers feed on lizards when they are young, graduating to larger mammals as they grow larger.

Although many snakes fall into this pattern of having relatively broad diets, which change from place to place and with the size of the snake, other snakes feed exclusively on a single type of prey throughout their lives. Some are very highly specialized indeed. The Australian bandy-bandy *Vermicella annulata*, a brightly black-and-white-banded burrowing snake, feeds only on blindsnakes of the genus *Ramphotyphlops*; it tracks its prey by scent and often eats blindsnakes as large as itself. Several other snake species specialize on lizard eggs; the lizard nests are presumably located by scent, and the snakes' teeth are modified so that each egg shell is slit as it is swallowed. Similar tooth modifications have developed in a variety of unrelated elapid and colubrid snakes that eat the eggs of reptiles as their primary source of food.

Other snakes take the much larger eggs of birds, and the African egg-eating snakes of the genus *Dasypeltis* are specialized colubrids with almost no teeth—only a few tiny ones in the very rear of the lower jaw. Where other snakes have teeth, these egg-eaters have a series of thick folds of gum tissue arranged in accordion-like folds. These folds act as suction cups on the smooth surface of the egg. After the snake swallows the egg—a remarkable feat in itself—it bends its neck sharply so that the egg is pushed against a series of sharp, downward-projecting spines that pierce the shell. These spines are formed by elongated projections from the snake's backbone. The egg's contents then flow down into the snake's stomach, owing to a set of special muscles that close the throat forward of the egg. As the egg

empties and travels further down the snake's throat, the blunt ends of some vertebrae crush the egg, forming a compact bundle of empty shell that can be easily regurgitated when the forward throat muscles have relaxed.

Many sea snakes are highly specialized feeders. Some sea kraits (*Laticauda* species) feed primarily on eels. Other sea snakes (some *Hydrophis* species) have slender forebodies and tiny heads that enable them to reach deep into crevices to obtain their prey. The turtle-headed sea snakes (*Emydocephalus* species) have vestigial teeth and feed only on the eggs of bottom-spawning marine gobies (a type of fish), using an enlarged, spade-like lip scale to scrape the eggs of coral or rock.

▲ *Steadying a bird's egg within its coils, an egg-eating snake Dasypeltis scabra begins the arduous task of swallowing its food. Possessing few teeth, this snake uses folds of gum tissue to grip the egg as it is swallowed.*

▼ *Almost completely engulfed, the egg will soon be pierced by projections on the snake's backbone and the shell regurgitated.*

▲ *By flicking its tongue a South American colubrid snake (genus Tachimenis) detects the scent of a tree frog and stalks to within striking distance.*

SENSE ORGANS

Snakes rely on a variety of sense organs to find their prey. Scent is probably the most generally important, and the forked tongue of snakes has been beautifully fashioned to gather information about chemicals in the environment. The two tongue-tips are widely separated. The tongue is in constant motion, regularly extruded from the mouth to sample particles in the air, the water, or on the ground, then withdrawing into the mouth to bring these particles to Jacobson's organ in the roof of the mouth. There the chemicals are analyzed, giving the snake accurate information about the presence of predators or prey in its local environment. In this way, Jacobson's organ has a similar function to the taste and smell organs of humans.

Vision is also important. The eyes of snakes differ considerably from those of other vertebrates and even from those of other reptiles such as lizards. For example, lizards focus their eyes by distorting the lens to change its radius of curvature, whereas snakes focus by moving the lens in relation to the retina. There are also several distinctive features in the eyes of snakes suggesting that their original ancestors may have had greatly reduced eyes—perhaps because they were burrowing creatures. When snakes later adopted life above ground again, larger eyes and better vision re-evolved but not in exactly the same way as before. The eyes of snakes are covered by transparent caps rather than eyelids (this also occurs in some types of lizards), giving them the "unblinking stare" so often interpreted as a sign of malevolent intentions.

Snakes have traditionally been thought to be totally deaf, because they have no external ear openings, eardrums, tympanic cavities, or eustachian tubes. However, they are capable of detecting even faint vibrations through the ground or water, and recent research suggests that some snakes may actually be able to hear airborne sounds as well. The pit organs of boid snakes (pythons and boas) and pit vipers allow them to detect warm-blooded prey because of the slight temperature difference between the prey and its surroundings. In practice, snakes use a combination of all these different senses to find their preferred food. We still have much to learn about the way in which information from these diverse sensory inputs is combined and interpreted in the brains of snakes.

WORMSNAKES

Three separate families of snakes—Anomalepididae, Typhlopidae, and Leptotyphlopidae, are often called "wormsnakes" because they resemble worms in size and general appearance. For example, their eyes are reduced to small, darkly pigmented spots which can tell the difference between light and dark, but probably little else, and their bodies are smoothly cylindrical. The blunt head merges smoothly with the rest of the body, and the tail is short and tipped with a small spine used to anchor the snake so that it can move forward more easily as it burrows through the soil. All are burrowers. In size they range from less than 10 centimeters (4 inches) long as adults, up to heavy-bodied snakes almost 80 centimeters (32 inches) in length. The three families are anatomically more primitive (that is, more like ancestral snakes known from fossils) than other living snakes. For example, most of them retain traces of a pelvic girdle, suggesting the presence of limbs at one time in their history. Their small size means that they do not leave a good fossil record, and this group probably arose much earlier than the oldest wormsnake fossils yet found, which are from the Eocene, about 50 million years ago. Some wormsnakes are very slender, like the 64 species of "threadsnakes" (family Leptotyphlopidae) of the southern United States, the West Indies, Central America, Africa, Arabia, and Pakistan. The 20 species of "blind wormsnakes" (family Anomalepididae) are found in continental Central and South America. The third family (Typhlopidae), known as "blindsnakes", is more diverse and contains about 150 species; most of these are found in Africa, Asia, and Australia but some species also occur in Central America, and one species (discussed later) is found almost worldwide. Although they differ considerably in some anatomical features—for example, threadsnakes have teeth only on their lower jaws, and blindsnakes only on their upper jaws—the general similarity between members of these three families suggests that they are closely related. All are non-venomous, feeding on soft-bodied invertebrates such as worms, or the eggs and larvae of ants and termites.

Wormsnakes rely on scent, rather than their rudimentary eyes, to locate their food. They are adept trail-followers, flicking their tongues in and out to pick up any faint chemical traces left by foraging ants, and analyzing these chemicals with the Jacobson's organ in the roof of the mouth. They can then follow these trails back to their source and find the ant brood. But how can a tiny snake enter an ant colony and defend itself against the bites and stings of the worker ants trying to protect the brood? The Australian

blindsnakes (*Ramphotyphlops* species) manage this simply by being reasonably large and having thick, smooth scales, so that the ants can't get a good enough grip to bite them. The Central American threadsnakes (*Leptotyphlops* species) are smaller, and they apparently repel the attacking ants by secreting repellant chemicals, writhing around to smear this secretion all over their bodies. It is such an effective technique that these tiny snakes can actually join hordes of marauding "army ants" as they travel through the forest, the snakes feeding on eggs and larvae from ant nests taken over by the army ants.

Most wormsnakes reproduce by egg-laying, and

▲ *The head of this giant blindsnake* Typhlops schlegelii *of South Africa shows several adaptations for burrowing. Enlarged scales form shields to protect the shovel-shaped snout and to cover the greatly reduced eyes, which are virtually useless underground.*

◄ *With their blunt heads, small cylindrical bodies, and rudimentary eyes it is easy to see how blindsnakes, such as* Ramphotyphlops nigrescens, *came to be known as wormsnakes.*

Haroldo Palo Jr/NHPA

Michael and Patricia Fogden

Esther Beaton/AUSCAPE International

EGGS OR BABIES?

Most snakes reproduce by laying eggs, but some species have evolved a different system. The developing eggs are retained within the mother's body instead of being laid in a nest, so that the young snake does not have to face the world until it is fully formed and ready for an independent life. This evolutionary transition from egg-laying (oviparity) to live-bearing (viviparity) has occurred at least 30 times in the ancestors of living snakes.

In some cases an entire group is live-bearing (like the filesnakes, family Acrochordidae), whereas in others a single group contains both egg-laying and live-bearing members. For example, within the family Boidae all of the boas are live-bearers, whereas all of the pythons are egg-layers. In a few cases, both types of reproduction occur within closely related species. The European smooth snakes, genus *Coronella* (family Colubridae) offer a good example. The two species in this genus are similar in most respects, and their geographic ranges overlap considerably. Nonetheless, the southern smooth snake *C. girondica* lays eggs, whereas the more northern species *C. austriaca* bears live young.

Why has live-bearing evolved in so many types of snakes? The geographic distributions of live-bearers give us a clue. Live-bearers are mostly found in climates cooler than their egg-laying relatives, and they are the only species to penetrate into severely cold areas. It seems that soils in these areas are too cold to allow successful incubation of eggs laid in the ground, whereas eggs retained within the mother's body can be kept much warmer because she can bask in the sun and select warmer shelters. Pregnant females of many live-bearing species spend most of their time basking, and this seems to accelerate the development of the eggs so that birth can occur while temperatures

are still favorable for activity and the young snakes can find safety for the winter.

If viviparity confers such advantages, why aren't all snakes live-bearers? The answer is that female snakes that retain developing young suffer *disadvantages* as well: they are physically slowed down by the volume of the litter they carry around, and they usually do not feed during most of their pregnancy. Females of egg-laying species do not suffer these "costs" for as long, and therefore may be less vulnerable to predators and be able to begin feeding again (and perhaps, laying another clutch of eggs) much sooner than the live-bearers.

▲ Splitting the shell membranes, baby sand racers Psammophis sibilans hatch from their leathery eggs.

◀ A female hog-nosed viper Porthidium nasutus gives birth, its offspring already on the alert as it emerges from the birth membrane.

females of one small American threadsnake, *Leptotyphlops dulcis*, may coil around their eggs and stay with them until hatching. It is difficult to imagine what benefits maternal attendance could offer these tiny snakes, but perhaps the female discourages small invertebrates that might otherwise prey on her eggs. Production of live young (viviparity) has been reported in some African blindsnakes (genus *Typhlops*).

Perhaps the most remarkable aspect of the reproductive biology of blindsnakes comes from the tiny flowerpot snake *Ramphotyphlops braminus*. This species is one of the smallest of all snakes (less than 15 centimeters, or 6 inches, as an adult) and has the broadest geographic range of any snake, including many small isolated islands in the Pacific Ocean. This huge range almost certainly results from the snake being accidentally spread around the globe by humans, who unwittingly transfer it in small containers of soil such as flowerpots (hence its common name). Like all wormsnakes, the flowerpot snake is completely harmless, although it has been reported to cause some problems in India by crawling into the ears of people sleeping on the ground. Its reproductive biology is quite bizarre, because all flowerpot snakes are females. They are triploid (that is, they have three sets of each chromosome in each cell, rather than two as in most animals) and reproduce parthenogenetically, with females giving birth to daughters who produce daughters who produce daughters, and so on, without any genetic contribution from a male. The same phenomenon occurs in several groups of lizards and some salamanders. Ultimately this reproductive system is likely to be an evolutionary dead end, because a parthenogenetic species lacks the genetic variation brought about by sexual reproduction. In the shorter term, however, it has obviously been a very sucessful strategy for these miniature snakes.

PIPE SNAKES AND THEIR RELATIVES

This group is a confusing mixture of medium-sized fossorial (burrowing) and semi-fossorial species, whose evolutionary relationships are obscure. They have a number of primitive features, such as a large left lung and traces of a pelvic girdle, which are lost in the so-called "advanced" snakes.

The pipe snakes, or aniliids (family Aniliidae), include one species of *Anilius* in northern South America, but there is considerable disagreement on what other snakes should also be included in the family. Some scientists would include *Anomochilus weberi* from Sumatra in the family, whereas others believe that it belongs in a separate family, Anomochelidae. Another unusual snake, *Loxocemus bicolor* of southern Mexico and Central America, is placed in the family Loxocemidae and may also be related to the pipe snakes; it burrows

in rotting foliage and loose earth and digs shallow tunnels in the ground, but is not exclusively burrowing. Seven *Cylindrophis* species in Southeast Asia (family Cylindrophidae) may be closely related to another group of Asian burrowing snakes, the shield-tailed snakes (family Uropeltidae). The genus *Cylindrophis* includes both egg-laying and live-bearing species, whereas all of the shield-tailed snakes are probably live-bearers.

Burrowing pipe snakes and shield-tailed snakes use their head to force their way through the soil, and the bones of their skull are solidly united, unlike those of most other snakes. A second unusual feature in pipe snakes, but one shared with boid snakes (described later), is the presence of rudimentary hind limbs, in the form of small spurs on either side of the vent. The shield-tailed snakes are burrowers and take their common name from the greatly enlarged scale near the tip of the tail. The scale may be compressed from side to side (in a few species it resembles a

▼ *Two pipe snakes: the blotched pipe snake* Cylindrophus maculatus *(below, family Cylindrophidae) of Sri Lanka, has a defense display which involves hiding its head, then flattening and raising its brightly patterned tail in mimicry of a cobra's head and hood. The relationship of the South American coral pipe snake* Anilius scytale *(bottom, family Aniliidae) to other pipe snakes is often debated. Its bright color pattern is in apparent mimicry of venomous coral snakes. However, its main defense tactic is to hide its head and present the blunt tail as an alternative "head".*

▲ *The greatest radiation of python species has occurred in Australia and the most widespread of these is the carpet python Morelia spilota. Numerous color varieties exist, including attractive black and yellow specimens from tropical north-east Queensland.*

diagonally-cut end of a salami sausage) and bears several small ridges or small spines, which hold a clump of soil. The soil blocks the tunnel behind the snake, protecting it from predators (including pipe snakes) which might otherwise capture and eat the shield-tail. The spectacular iridescent colors of the shield-tails (and many other types of snakes that burrow in wet soil) are not due to pigments in the skin, but to the microscopic structure of the scales. These bear small ridges that reduce friction against the surrounding soil while the snake is burrowing, and incidentally diffract light, producing an attractive iridescence with all the colors of the rainbow.

SUNBEAM SNAKE

The sunbeam snake *Xenopeltis unicolor*, a medium-sized ground-dwelling snake of India and Southeast Asia, seems to be only distantly related to other living snakes and is usually placed in a family by itself, the Xenopeltidae. Vestiges of a pelvic girdle and rudimentary hind limbs are primitive features it shares with the pipe snakes, although this may not indicate close relationship. It grows to about 1 meter (3¼ feet) in length and has a round body and glossy scales like those of the shield-tails, presumably for the same reasons. One very unusual feature is that the teeth of the lower jaw are set on a loosely hinged dentary bone. Surprisingly, the sunbeam snake doesn't seem to show any corresponding peculiarities in its diet, because it feeds on a wide variety of vertebrates including other snakes, frogs, and rodents (mice and rats).

PYTHONS, BOAS, AND THEIR RELATIVES

The 60 species of boid snakes (family Boidae) include the largest of all living snakes, but also many smaller species—for example, the pygmy python *Antaresia perthensis* of the western Australian deserts grows to only 60 centimeters

(24 inches), and feeds mostly on geckos and small mice. Nonetheless, it is the larger species that attract the most public interest. The record for the "longest snake in the world" goes to either the reticulated python *Python reticulatus* of Asia, reliably measured up to 10 meters (almost 33 feet), or the semi-aquatic anaconda *Eunectes murinus* of South America, also measured up to at least 10 meters and much more heavy-bodied than the reticulated python. Other giants include the Indian python *Python molurus,* to 9 meters (29½ feet), the African python *P. sebae,* 8 meters (26½ feet), and the Australian scrub python *Morelia amethistina,* more than 7 meters (23 feet).

Many authorities believe that the snakes usually grouped together as "boids" are actually a combination of different types that are only distantly related to each other. The wood snakes (usually considered a separate family, the Tropidophiidae) are found from Mexico to northern South America, and on many offshore islands as well. These 20 species of small ground-dwelling boa-like snakes seem to be the living representatives of a very ancient lineage, with fossils known from the Paleocene, about 60 million years ago. Their closest relatives may be the two small species of snakes *Bolyeria multocarinata* and *Casarea dusumieri* (family Bolyeriidae), found only on tiny Round Island in the Indian Ocean, although fossils have also been found on nearby Mauritius. There has been massive environmental degradation on Round Island, mostly because of grazing by introduced pigs, which has destroyed much of the habitat available to the boas on the island. Their extinction seems almost inevitable.

The two other groups of boids both contain species that attain very large body sizes. Pythons (subfamily Pythoninae) are found mostly in the Old World and consist entirely of egg-laying species. In contrast, boas (subfamily Boinae) are mostly in the New World and are viviparous

◄ Best known of the large snakes, the boa constrictor Boa constrictor is by no means the largest, and is dwarfed by several other boids. Commonly thought of as a jungle snake, the numerous subspecies are found in a variety of habitats from semi-desert to rainforest.

species—that is, the female retains the developing eggs within her oviduct, instead of laying them, and gives birth to fully formed offspring.

Many people have a mental picture of a python as a large, heavy-bodied snake lying on a branch in a tropical forest, waiting to drop onto some unlucky explorer. The truth of the matter is quite

different. Many pythons are quite small and do not live in tropical forests at all. Indeed, some are burrowers living in desert sands. In fact, although they have representatives in Africa and Asia, true pythons are most diverse in Australia. While it is true that all pythons are non-venomous, their feeding habits vary from species to species. Some

▼ The blood python Python curtus is a short, stout python of Southeast Asia, where it inhabits swamps, marshes, and rainforest streams. Its common name is derived from the deep red coloring of some individuals.

feed on frogs, some on lizards, some on mammals, some on birds. They all rely on constriction to subdue their prey, throwing a series of coils around the animal as soon as it is seized. These coils suffocate the prey by tightening every time it exhales and preventing it from drawing another breath—although stories of prey animals being crushed to jelly, and of all their bones being broken by constriction, are simply untrue. One desert species, the woma python *Aspidites ramsayi* of central Australia, uses an interesting variation to conventional constriction. This species catches many of its prey down burrows, where there isn't enough room for it to throw coils around the prey. Instead, the woma just pushes a loop of its body against the unlucky mammal so that it is squeezed (and soon suffocated) between the snake and the side of the burrow. Unfortunately for the woma, this technique doesn't immobilize the prey as quickly as would "normal" constriction, so many adult woma pythons are covered in scars from retaliating rodents.

All pythons are egg-layers, like most other reptiles, but pythons are unusual among reptiles in the care they afford their developing eggs. The female may build a "nest" of vegetation, by coiling under loose leaf-litter, for example, or simply selecting an appropriate well-insulated burrow. She coils around the clutch after laying and remains with her eggs until they are ready to hatch. She may leave to bask in the sun or to drink but will not feed until after her maternal duties are complete. Although pythons are generally ectothermic like other reptiles—that is, they rely on heat from the environment to keep themselves warm—brooding female pythons actually generate heat by shivering, and this keeps the eggs at a high and stable temperature throughout development. It is very expensive in terms of the female's own energy reserves: she may lose up to half her own body weight between egg-laying and the end of incubation, and it may take her two or three years to regain enough energy reserves to breed again. However, it means that the eggs develop rapidly to hatching, and are safe when the air temperature is low. Perhaps for this reason, some pythons can reproduce successfully even in relatively cold areas. Like all other snakes, pythons do not take care of their offspring after hatching; their maternal responsibilities finish when the young snakes emerge from the eggs.

Boas are similar in many ways to the pythons,

▼ *Loosening its coils after constricting a rat, a Burmese python* Python molurus bivittatus *repositions the prey to begin swallowing. One of the largest snakes, the rapid growth rate and attractive patterning of this python have resulted in its popularity in both the leather and live animal trades.*

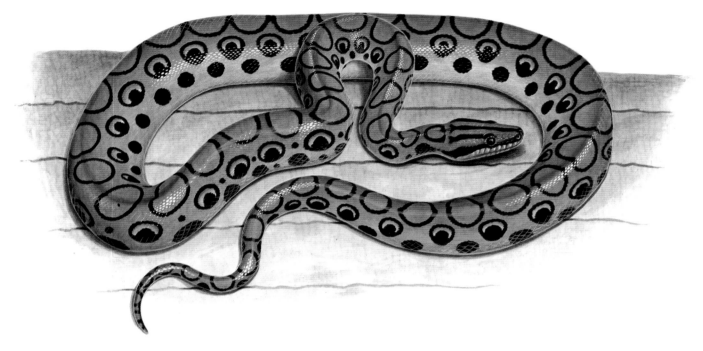

except that they bear live young instead of laying eggs. The best-known boas may be large species, but there are also smaller types. They are mostly found in Central and South America, although two genera occur in North America, the rubber boa *Charina bottae* and the rosy boa *Lichanura trivirgata*, both relatively small and secretive. Small groups have also made their way somehow to Madagascar (*Sanzinia* species) and to New Guinea and nearby Pacific islands (*Candoia* species); their ancestors may have rafted across from Central America, like the ancestors of the Pacific iguana lizards—or perhaps their descendants rafted the other way. The anaconda *Eunectes murinus* is a giant semi-aquatic snake of South American rivers, where it spends much of its time lying in wait at the water's edge for unwary mammals and caiman, a type of crocodilian. Probably the best-known species is the *Boa constrictor*, a medium-sized boa at 4.5 meters (14¾ feet), distributed from northern Mexico to Paraguay and Argentina, and often kept as a pet. The danger posed by large boid snakes is undoubtedly grossly exaggerated, but there are reliable reports of predation on humans (usually children) by unusually large African pythons, Indian pythons, reticulated pythons, and anacondas.

FILESNAKES

In many respects, the three totally aquatic species in this family (the Acrochordidae) are among the most unusual of snakes. Their skin is loose and baggy and looks as if it is one or two sizes too large for the snake. The skin is covered with small granular scales like the surface of a file, giving these snakes their common name. They have lost the enlarged ventral scales, the belly-shields that characterize most terrestrial snakes, and have trouble moving around on land. Another name

that describes them well is "elephant's trunk snake", applied to a freshwater Asian species, *Acrochordus javanicus*. The Arafura filesnake *A. arafurae* lives in rivers and freshwater billa-bongs in Australia and New Guinea, and the third species, the little filesnake *A. granulatus*, inhabits estuarine and coastal areas of the Indo-Pacific.

When you see a filesnake underwater, its strange skin begins to make sense. The loose skin flattens out ventrally as the snake swims, giving it a flattened profile like a sea snake and thus enabling it to swim more efficiently. Its rough skin is used to capture prey in a rather surprising way. Filesnakes feed on fishes, and constrict them in the same way that pythons constrict their prey on land. Squeezing a slippery fish would be impossible without small roughened scales that

▲ *Named for the iridescent sheen of its scales, the Brazilian rainbow boa* Epicrates cenchria cenchria *(top) feeds on the ground and is often found near village outskirts, where there is a steady supply of rodents. Sand boas are unusual boids in that they are exclusively Old World in distribution and adapted for a burrowing lifestyle. Illustrated is the Kenyan sand boa* Eryx colubrinus loveridgei *(above), the southern-most representative of this group.*

Jean-Paul Ferrero/AUSCAPE International

COLUBRID SNAKES

Most living snakes belong to a single family, the Colubridae, often called the "harmless snakes", even though a few of them have very toxic venom. About 1,600 species are recognized, and they occur on all continents except Antarctica (which has no snakes at all). This remarkable assemblage is extraordinarily diverse, but apparently fairly recent in geological terms. The earliest definite colubrid fossils come from the Oligocene, about 30 million years ago, and the colubrids are one of the most spectacular success stories in the world's evolutionary history since that time. They are the most common snakes almost everywhere snakes occur, with the notable exception of Australia. They have lost all trace of the pelvic girdle, and in most species the skull is highly modified so that it is very flexible, allowing large prey items to be swallowed.

Venom has evolved independently in several different groups of snakes. Colubrid venom is really just modified salivary secretion from Duvernoy's gland, in the upper jaws. Because saliva contains components that break down tissue and begin the process of digestion, it's easy to imagine the evolutionary pathway leading to the appearance of venom in snakes. All that is required in the early stages is that some toxic saliva trickles down into the prey item as it is held in the snake's mouth. Snakes with more toxic saliva, or with larger teeth so that the saliva could penetrate the prey more easily, had an advantage because their prey were killed more quickly. Over many generations, natural selection favored snakes that had more and more toxic venom, and larger and more elaborate teeth with which to deliver the venom to the prey.

The earliest venomous snakes may have been "rear-fanged" (opisthoglyphous)—that is, their enlarged teeth (fangs) were at the rear of their mouth—and this is the type of system seen in the venomous colubrid snakes today. In bio-mechanical terms, the rear of the mouth is where the tooth can exert the greatest force on a prey item; many non-venomous snakes have enlarged teeth at the rear of their mouths, which they use for slitting relatively hard objects (like eggs) or puncturing prey animals that inflate themselves with air (like toads) and hence may be difficult to ingest. Most snakes subdue and swallow their prey without venom, but several groups have evolved toxic secretions to kill prey more rapidly and perhaps to begin the process of digestion before the prey item reaches the stomach. Among the colubrids, potent venoms have arisen in a number of species, perhaps the best known being three arboreal African snakes; the boomslang *Dispholidus typus* and the two species of vine snakes, genus *Thelotornis*. Snakes of both these genera have caused human fatalities, including famous herpetologists who underrated their danger.

▲ *An underwater close-up of an Arafura filesnake* Acrochordus arafurae *reveals the coarse textured skin with which it holds its slippery fish prey.*

▼ *Gaping in threat, a toad-eater* Xenodon rabdocephalus *displays the enlarged rear teeth with which it punctures toads and frogs.*

Michael and Patricia Fogden

can penetrate the slime covering the fish's scales. Filesnakes are unusual in behavior, ecology, and physiology as well as body form and structure. They seem to be specialists in coping with low rates of energy availability: they eat rarely, have little capacity for sustained exercise, and females may reproduce only once every few years. Still, they are able to acquire energy fairly rapidly if food does become available, and can occur in remarkably high numbers in suitable areas. In tropical Australia, Aboriginal people harvest many of these snakes just before the beginning of the annual monsoonal rains, when water levels are at their lowest. Aborigines catch the snakes by groping around blindly in the muddy water, recognizing the snakes by the distinctive feel of their rough skin. They are cooked by the simple technique of throwing them on hot campfire coals, and then eaten in their entirety.

Marine filesnakes are also harvested at sea by humans, but this is a commercial industry based on the value of their skins for clothing and fashion apparel.

◀ A rear-fanged colubrid, the long-nosed tree snake Ahaetulla nasuta has grooves in front of its horizontally-pupiled eyes to allow unobstructed forward vision for hunting.

Mike Price/Bruce Coleman Limited

HOW OFTEN DO SNAKES REPRODUCE?

In tropical areas where conditions are suitable for reproduction year-round, females of some egg-laying species may produce more than one clutch of eggs in a single year. However, highly seasonal reproductive cycles are common even in tropical snakes. This is probably because most tropical areas actually show significant seasonal variation in characteristics such as rainfall, which may influence the food supply for hatchlings. In temperate-zone habitats, snakes generally produce only a single clutch or litter per year, because winter temperatures are too low for reproduction (and often, for any activity at all).

In many kinds of snakes, adult females don't reproduce every year: instead, a female is likely to skip opportunities for reproduction, perhaps producing offspring only every second or third year. These low reproductive frequencies seem to result from the female's difficulty in accumulating the amount of energy needed to produce a full clutch every year. Particularly in live-bearing species living in cold areas (where the activity season may be only a few months long), a female may have little time to feed between the time she gives birth and the time she must enter her winter retreat. For this reason, populations of geographically widespread species may vary in the frequency at which the females reproduce: females in colder areas may be able to reproduce only once every few years, whereas females of the same species living in warmer areas can repro-duce every year. This pattern is seen in several species, including both the European viperids and the Australian elapids.

Females may also skip a year between successive reproductions if food supply is limited. In the Saharan desert viper Echis colorata, some females living near oases (where food is plentiful) reproduce every year, whereas those in drier areas do not. In the European adder and the Australian water python Liasis fuscus, few females reproduce during years when rodent populations are low and food is therefore hard to find. Female Arafura filesnakes have very low reproductive frequencies, with an average of only 10 percent of females reproducing each year. The only years of significant reproduction in this species come after exceptionally prolonged flooding, when the snakes can capture many fishes in shallowly flooded areas. In all of these species, however, males tend to mate every year, because they do not need to gather enough energy to produce a large clutch of eggs or young.

▼ The considerable energetic cost of producing relatively large eggs, such as those being laid by this yellow-faced whip snake Demansia psammophis, is a major factor determining the frequency of reproduction in snakes.

Pavel German

Some authorities believe that the family Colubridae is such a huge group that it should be divided into many smaller lineages, but there are still difficulties in identifying relationships. Recent studies using a combination of biochemical and morphological (body structure) techniques have identified some groupings within the Colubridae, but have left many uncertainties; for example, the "burrowing vipers" of the genus *Atractaspis* found in Africa and the Middle East are very distinctive, with no clear relationships to other kinds of snakes. So let us begin with the groups within the Colubridae that do seem to represent natural evolutionary lineages.

Subfamily Homalopsinae

The homalopsines are an array of about 40 rear-fanged aquatic snakes of Asian and Australian waters. One land-dwelling genus, *Brachyorrhos*, probably belongs here too. All are live-bearers, and many are mangrove-dwellers. Some species are very distinctive, like the tentacled snake *Erpeton tentaculum*, so-called because of the strange protuberances on its head. Most of the homalopsines are fish-eaters, but the white-bellied mangrove snake *Fordonia leucobalia* is a specialist on crustaceans, especially crabs. It stalks them at night on mangrove mudflats left by the receding tide. The hard shell of a crab makes it difficult to seize, so the snake uses a special technique. When the snake is close enough, it launches a strike, not *at* the crab, but *above* it. By striking above the crab, the snake's forebody pushes the crab down into the soft mud, and the snake can then turn and bite it more carefully to introduce venom. This seems to stop the crab's struggles quickly, and the snake then proceeds to eat the unfortunate crustacean—or, if the crab is too big, to remove and eat the legs only. This is probably the only snake that can actually tear pieces off a prey item (because the crab sheds its legs quite readily); all others must eat the prey entire.

Subfamily Xenodermatinae

The xenodermatine colubrids of India and Southeast Asia are unusual in having upturned edges on the scales bordering the lip, and expanded bony plates on the spines of the vertebrae. *Xenodermus javanicus* is a frog-eating species found in moist soft earth in wet cultivated areas, such as dykes between rice fields, in parts of Asia.

Subfamily Calamariinae

The calamariines or "dwarf snakes" of East Asia form another clearly differentiated colubrid group, possibly related to the xenodermatines. There are about 80 species, all of them small snakes that seem to feed mostly on earthworms and insects. These small secretive snakes are slow-moving and relatively defenseless, and hence are often eaten by other snakes.

Subfamily Pareatinae

One Southeast Asian group, the pareatines, are particularly interesting because of their specialized diet. Like a distantly related genus of colubrids from tropical Central and South America, *Dipsas*, they feed on snails and have evolved some remarkable modifications of their body structure and behavior to suit them to this unusual diet. The lower jaw is strengthened by the fusion of adjacent scales and can be inserted into the opening of a snail's shell; the long front teeth then hook the snail's body and pull it out with twisting movements. The snake does not consume the snail's shell, which it probably could not digest anyway.

▶ *A puzzle to taxonomists, the spotted harlequin snake* Homoroselaps lacteus *of southern Africa is venomous and has fixed front fangs like an elapid and yet has been classified with both the vipers and the colubrids. Often found in termite mounds, this species feeds on blind snakes and legless lizards.*

Subfamily Boodontinae

The boodontine colubrids of Africa and Madagascar are a very large group, including both harmless species and rear-fanged species. Some are aquatic fish-eaters, some are terrestrial with broad diets, some feed mostly on mammals, some on lizards, some specialize on eating other snakes, and one genus (*Duberria*) specializes on slugs. Both egg-laying and live-bearing occur within the group, and even within a single genus (*Aparallactus*). Several of the species within this large and diverse group are known as "house snakes" because they often enter houses to feed. The western keeled snake *Pythonodipsas carinata* is an unusual-looking species that closely resembles horned vipers found in the same area; this mimicry may confuse predators and hence reduce the keeled snake's vulnerability to them.

Stephen Dalton/NHPA

Subfamily Natricinae

One of the most successful colubrid groups in North America, Asia, and Europe is the subfamily Natricinae: garter snakes and their relatives. This group probably had its origin somewhere in the Old World, possibly in Asia, but has spread widely through the New World as well. The Old World forms are mostly egg-layers, like the common grass snake *Natrix natrix*, which is common over much of Europe, even at high latitudes. Egg-laying snakes are usually not able to survive severely cold climatic regions, because they require high soil temperatures for their eggs to develop. Female grass snakes overcome this problem by migrating long distances to find

▲ The most widespread European snake, the grass snake *Natrix natrix* is associated with water throughout much of its range.

▼ Like its close relatives in the genus Dipsas, the snail-eater *Sibon annulata* of Central America has a specialized diet of snails.

Michael and Patricia Fogden

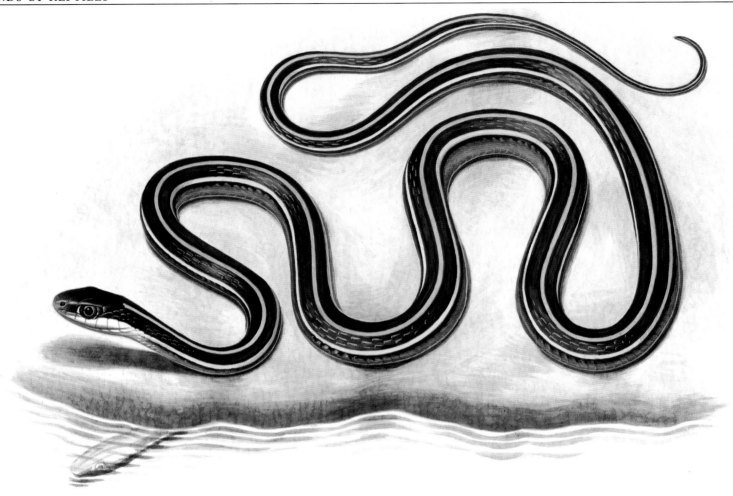

▲ *A slender version of the closely related garter snakes, the eastern ribbon snake* Thamnophis sauritus sauritus *is a typical New World natricine snake in that it is live-bearing and semi-aquatic, rarely being found far from water.*

suitable incubation sites, but even so, such sites are scarce in areas like northern Sweden. How, then, can grass snakes survive in these areas? They do so by taking advantage of manure heaps on farms. These huge piles of cow manure are heated by the action of microbes, and snakes from a wide area may converge onto a single farm and lay thousands of eggs in the same manure pile. In this way, their eggs are kept at high-enough temperatures to complete development and to hatch before the first frosts of winter. In natural conditions, before manure piles were available, the snakes probably nested under large, flat sun-warmed rocks on south-facing slopes because these would have offered the only nest sites warm enough to allow for successful embryonic development.

Not all natricines are egg-layers, however. One semi-aquatic Asian species, *Sinonatrix annularis*, is viviparous, as are all of the New World natricines such as garter snakes and watersnakes. The North American natricines are a spectacularly diverse group and remarkably common in many areas throughout their range. In some severely cold areas in south–central Canada, red-sided garter snakes *Thamnophis sirtalis parietalis* gather together in huge groups to spend winter in the few sites where underground crevices are deep enough for them to escape the winter freeze. These small

snakes may travel many kilometers from their hibernation sites to their summer feeding grounds, the frog-rich swamps and meadows. When the first warm days of spring bring the hibernating snakes out to bask at the entrance to the den, the numbers of snakes to be seen are nothing short of astounding. Males emerge first and wait by the den entrance for the females, which are larger. Reproductively receptive females secrete a special chemical substance, a pheromone, which stimulates mating activity by the males. Because of the huge numbers of snakes present, a female may be surrounded by a writhing ball of dozens of amorous suitors. It seems to be a case of "first come, first served", but some males have an advantage because they also produce the female pheromone or something very like it. This confuses the other males, who don't know who to try to mate with, and as a result these "she-males" are more successful than normal males at mating.

Because garter snakes and watersnakes are so common and so widely distributed in North America, they are among the best-known snakes in many respects. For example, we know a great deal about their feeding habits. Different species of garter snakes found in the same area tend to eat different things, although it is not clear whether this is because of competition between species. Scientists can find out the food preferences of a newborn garter snake by testing its reaction to different prey odors presented to the snake on cotton swabs. It turns out that prey preferences seem to be genetically programed and that they differ between species of garter snakes, and even between different populations of the same species.

Natural selection has also modified the snakes' tolerance to toxins in its prey. For example, garter snakes living in areas where toxic newts are common are very resistant to the poisons produced by the newts. In areas where the newts don't occur, or where they are not as toxic, the local garter snakes don't show this resistance.

Subfamilies Xenodontinae and Pseudoxenodontinae

The xenodontine colubrids of the temperate and tropical New World, including the West Indies and the Galapagos, are less well known ecologically than the other members of the colubrid family. Recent research indicates that the West Indies xenodontines feed mostly on anoline lizards ("false chameleons") and to a lesser extent on frogs, whereas boas and vipers living in the same area eat mainly birds and mammals. The North American hognose snakes (genus

Heterodon) feed primarily on toads and use their broadened noses for digging their prey out of burrows. The unusual nose is also used in a remarkable defensive display. These heavily-built snakes look rather like vipers, and the resemblance is strengthened by the way they flatten their head and neck when harassed, and hiss and strike vigorously. If this formidable display fails to deter the attacker, the snake adopts a very different strategy: turning over on its back and feigning death.

The pseudoxenodontine colubrids are an Asian group of small to medium-sized snakes that feed mostly on frogs and toads. When attacked, they show a "death-feigning" display similar to that of the North American hognose snakes; the neck and forebody are flattened, the mouth is opened, the lips are drawn back, the tail is vibrated, and the snake rolls over onto its back.

Several different types of toad-eating snakes seem to have independently evolved this kind of dramatic display, and some researchers have suggested that physiological modifications for toad-eating may play some role in the behavior. Toad-eating snakes tend to have large adrenal glands, and perhaps are more likely to go into some kind of shock when attacked.

▼ *The western hognose snake* Heterodon nasicus *uses its shovel-shaped snout to dig for toads, its main prey. These are then dispatched with the snake's enlarged rear teeth.*

Subfamily Colubrinae

The colubrine subfamily is very diverse and wide-ranging, found over the entire range of the Colubridae family. The evolutionary relationships among this subfamily are particularly difficult to unravel. For example, although all of the North American natricines seem to result from a single ancestral group that came from Eurasia, this isn't true of the North American colubrines; several different migrations seem to have occurred, so that several different evolutionary lineages of colubrines may be present in North America. These include the spectacularly colorful king snakes and milk snakes (genus *Lampropeltis*), which have enormous variation in color even among individuals within a single population.

Colubrines have adapted to a wide variety of ecological niches, with the Indian wolf snake *Lycodon aulicus* often found in houses, where it preys on lizards and mice. The Asian "kukri snakes" of the genus *Oligodon* got their common name from the supposed resemblance of their enlarged rear teeth to the ceremonial dagger (kukri) used by local tribes. Kukri snakes feed mostly on the eggs of reptiles, and their teeth are modified to slit the shell as the egg is swallowed.

Many colubrines have enlarged rear teeth, with or without the development of significant venom. One of the groups with venom—quite toxic in some species—is a genus of slender, tree-dwelling snakes, *Boiga*, which includes some of the most spectacular colubrids. They are nocturnal foragers, remarkably adept climbers, and prey on birds and any other small vertebrate that they encounter. One species, the brown tree snake *B. irregularis*, is widely distributed through Australasia and was accidentally introduced to the tiny Pacific island of Guam after the Second World War, probably in

Jack Dermid/Bruce Coleman Limited

▲ *The scarlet king snake* Lampropeltis triangulum elapsoides, *a harmless colubrine, is one of several snakes whose banded patterns are believed to mimic the coloration of the highly venomous coral snakes. Like the king cobra, the "king" refers to its habit of eating other snakes.*

▶ *The Mandarin ratsnake* Elaphe mandarina *is a brilliantly colored colubrine of high altitude regions in China. Like other ratsnakes it is a constrictor and feeds on warm-blooded prey, particularly rodents, giving the group its common name.*

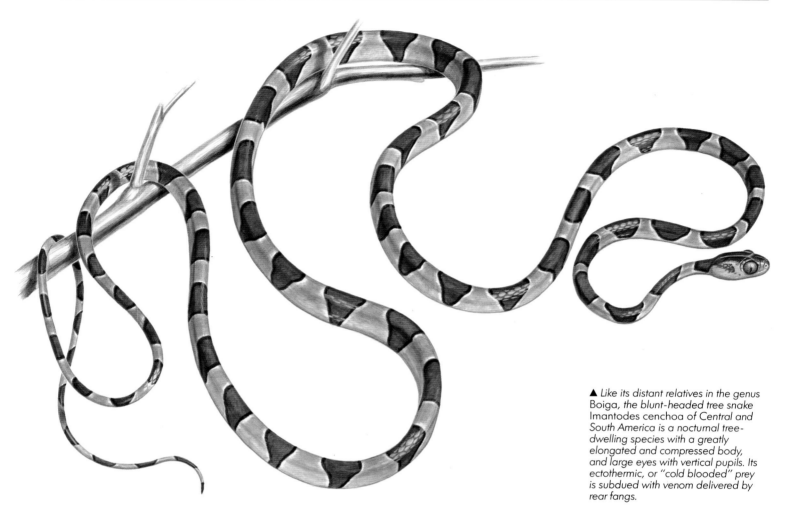

▲ *Like its distant relatives in the genus Boiga, the blunt-headed tree snake Imantodes cenchoa of Central and South America is a nocturnal tree-dwelling species with a greatly elongated and compressed body, and large eyes with vertical pupils. Its ectothermic, or "cold blooded" prey is subdued with venom delivered by rear fangs.*

military equipment brought back from the jungles of Guadalcanal and similar areas. Guam had no endemic snakes, but is (or was) home to an interesting array of bird species found nowhere else in the world. They had evolved on this snake-free island and so were easy prey to the depredations of the introduced tree snake. Within the past 50 years the tree snakes have increased enormously in numbers, and the native birds of Guam have been driven to the brink of extinction or beyond. Without the birds to control their numbers, the insects have increased unchecked, and the entire jungle ecosystem is under severe threat. This tragic example highlights not only the dangers of introducing "foreign" animals to new areas but also the important ecological role played by snakes. Presumably the evolution of various types of snakes and their dispersal into new areas over evolutionary history has profoundly influenced the nature of the ecosystems around us today.

FIXED-FRONT-FANGS: FAMILY ELAPIDAE

The dangerously venomous snakes belong to two different groups. Both first appear in the fossil record quite recently (about 20 million years ago, in the Miocene) and have fangs at the front of the mouth, and hence in a much better position to deliver venom with a rapid strike. One group (the proteroglyphs), which includes cobras and their relatives, consists of species with "fixed" fangs; the fangs are attached to the upper jawbone like normal teeth. This sets an upper limit to the size of the fangs, which must be small enough for the snake to be able to close its mouth without the downward-projecting fangs piercing the lower jaw and dragging along on the ground.

The other major group of highly venomous snakes, the vipers, have evolved an ingenious solution to this constraint. Their fangs are attached to a small bone that can rotate so that the fangs lie back along the length of the upper jaw when the snake's mouth is closed but can swing forward into striking position when the mouth is opened. Their dentition is known as solenoglyphous, meaning "pipe-tooth", in reference to the large hollow fangs of these snakes. Many vipers have enormously long fangs, whereas the fixed-front-fang group have relatively short fangs. For example, the fangs of the king cobra *Ophiophagus hannah* (the largest proteroglyph, which grows to more than 5 meters or 16½ feet) are not much larger than those of the adder, one of the smallest viperid snakes.

Most fixed-front-fang snakes, whether they live in the oceans or on the land, are long slender

▲ *Flourishing the trademark of the cobras, the most widely recognized elapids, an agitated black-necked spitting cobra* Naja nigricollis *spreads its hood in a defense display.*

Mark Boulton/Bruce Coleman Limited

animals that gather their food by actively searching for it. Many that live on land rely on scent to locate prey trails and follow them to the hiding-places of their prey. For example, many elapid snakes are small species that catch sleeping lizards in their nocturnal retreats. Although there are numerous exceptions, most of the fixed-front-fang snakes feed on ectothermic ("cold-blooded") prey such as fishes, frogs, lizards, and other snakes.

Traditionally fixed-front-fang snakes were assigned to two families—the terrestrial snakes such as cobras, kraits, mambas, taipans, and coral snakes in one family, and marine or sea snakes which spend all or part of their lives in the sea in the other family. A number of recent studies have shown this classification to be much too simplistic, and a number of new and conflicting classifications have been proposed. However, in all these classifications a number of distinct groups are recognized, and these are treated below (as subfamilies) under the umbrella of a single family: the Elapidae.

Terrestrial elapids are diverse and abundant on all tropical and subtropical landmasses. So far, 180 species of terrestrial elapids are known. Many elapids are famous for the danger they represent to humans. They include large and formidable species such as the black mamba *Dendroaspis polylepis*, the green mambas *D. angusticeps* and *D. viridis*, and a variety of other mambas and cobras

THE BIGGER THE BETTER

Females grow larger than males in most types of snakes. This is true of almost all boids, so the largest snake alive today is almost certainly a female, probably an anaconda in some South American river. Sometimes the disparity between the sexes can be extreme; for example, in one mating pair of Arafura filesnakes *Acrochordus arafurae* the male weighed only a tenth as much as his mate.

It is the general rule in the animal kingdom that the female is larger than the male, and Charles Darwin in the nineteenth century suggested an explanation for this phenomenon. The number of offspring that a female produces depends upon her own body size, so larger females have more eggs or babies. The range in clutch sizes can sometimes be quite wide, even within a single species; for example, in the carpet python *Morelia spilota* young females produce only about six eggs, whereas older larger females often have more than 30. This means that genes resulting in large body sizes for females are likely to increase the number of offspring a female produces during her lifetime, and thus be favored by natural selection. Over long periods of time, genes for large size in females should become more and more common in the population, and females are likely to grow larger than males unless there is some opposing selective force favoring even-larger size in males.

If we look at the kinds of snakes in which males do equal or exceed females in size, we find evidence that the relative size of the two sexes seems to depend on the mating system. Males tend to be larger than their mates mostly in species whose males engage in physical combat with each other during the mating season. This behavior differs from species to species, but generally consists of two males intertwining their

in Africa; cobras, king cobras and kraits in Asia; coral snakes in Central and South America; and taipans, brown snakes, death adders, and the like in Australia. However, even though it is the large species that attract popular attention, most elapids are actually small snakes that pose no threat to humans.

Australasian elapids

Australia is the only continent where most of the snakes are venomous, and it also has the snakes with the most toxic venom. Both of these attributes are due to elapid snakes. Australia, together with New Guinea and the island archipelago extending from its eastern edge, has long been isolated from the rest of the world, and during much of its geological history it was even

Hellio and Van Ingen/NHPA

bodies while each attempts to push the other's head downward. It seems to be a test of physical strength and one that is almost always won by the larger of the two males. Detailed studies of the adder *Vipera berus* confirm that winning fights like this increases a male's mating opportunities, so that—according to Darwin's hypothesis of sexual selection—we would expect male-male combat to favor the evolution of large body size in males. In snakes that do not engage in male-male combat, such as garter snakes *Thamnophis* or filesnakes *Acrochordus*, larger size does not seem to enhance a male's reproductive success.

Although there are many exceptions to the rule, the overall pattern fits this prediction. Males do tend to be at least as large as their mates in snake species with male-male combat.

David Curl

▲ In grass snakes Natrix natrix, as in most snakes, the male is smaller than the female. The dark-colored snake in this courting pair is the male.

◄ Two male Malagasy giant hognose snakes Lioheterodon madagascariensis *intertwine as they combat over access to females.*

more isolated as it floated northward after the breakup of the great southern continent, Gondwana. This separation meant that relatively few groups of plants and animals found their way to Australia. The successful ones faced little competition from other groups and thus have radiated very extensively. This is clearly true of eucalyptus trees and marsupials (kangaroos, koalas, etc.) but also true of various groups of amphibians and reptiles. It seems as though elapid snakes reached Australia just after it collided with Asia about 25 million years ago. The Australian elapids (subfamily Oxyuraninae) have diversified to fill a wide variety of ecological niches, with one genus (the death adders, *Acanthophis*) evolving to look and act remarkably like the vipers of other continents. The first

John Cancalosi/AUSCAPE International

◄ A stocky ambush predator, the death adder Acanthophis antarcticus *has evolved an appearance more like that of unrelated vipers than of its relatives in the family Elapidae. Like many vipers, the death adder can lure prey with its modified tail.*

scientists to describe Australian snakes believed that *Acanthophis* was actually a viper. Other Australian elapids have evolved to resemble colubrid "whipsnakes" and small banded burrowing snakes in other countries. This kind of convergent evolution happens when animals of different evolutionary backgrounds are exposed to similar environments and therefore similar evolutionary pressures. In the case of death adders, the important similarities with viperid snakes seem to be that these heavily-built elapids ambush their prey rather than searching actively for it like most of the other proteroglyphous snakes.

In Australia, elapids occur in a very wide variety of habitats. Small offshore islands often have distinctive populations of brown snakes or tiger snakes, and even on adjacent islands there may be major differences between snakes. For example, some island tiger snakes (genus *Notechis*) reach more than 2 meters (6½ feet) in length, whereas snakes on other islands do not attain even 1 meter (3¼ feet). Tenfold differences in average body weights of tiger snakes have been recorded from islands just a few kilometers apart. The reason for this size variation seems to be the food supply: tiger snakes grow large only on islands where large prey are available, and this

means only where there are large aggregations of nesting seabirds, especially muttonbirds. For most of the year food for snakes is scarce, and they survive by catching occasional small lizards; food is abundant for only a few weeks each summer, after the muttonbird chicks hatch and before they grow too large for the snakes to swallow. On islands without nesting colonies of muttonbirds, the tiger snakes can find only small prey and never grow very large.

Mambas

Africa has a variety of elapids ranging from small secretive creatures such as the shield-nosed snakes (genus *Aspidelaps*) to cobras and the renowned mambas (subfamily Dendroaspinae). The black mamba *Dendroaspis polylepis*, which can grow to 4 meters (13 feet) in length, is the most terrestrial of the four mamba species, and the subject of many horrifying tales. The animal's aggressiveness, speed, and venom toxicity are highly exaggerated in most accounts, but there is no doubt that these slender olive-brown elapids are among the most dangerous of all snakes. The threat display of an angry mamba—head and neck held high, mouth gaping open—is a truly terrifying sight.

▼ With its neck thrown into a distinctive "S" curve and its mouth agape, the defense pose of the eastern brown snake Pseudonaja textilis *is a warning wisely heeded. This fast-moving venomous snake is part of a successful species group found throughout Australia.*

Mike W. Gillam/AUSCAPE International

Dr Norman Myers/Bruce Coleman Limited

Cobras and their allies: subfamily Elapinae

Although cobras are distributed widely in Asia, they are most diverse in Africa, where there are 10 species: in the arid northeast, the Egyptian cobra *Naja haje*; in hot, humid western and central African jungles the forest cobra *N. melanoleuca*, the spitting cobra *N. nigricollis*; in the rocky fields and mountains of South Africa, the Cape cobra *N. nivea*; in the great freshwater lakes the water cobra *Boulengerina annulata* which feeds on fish, emerging from among lakeshore boulders at dusk to hunt for prey; and in southern Africa a small cobra, the rhinghals *Hemachatus haemachatus* only distantly related to the others. The rhinghals is distinctive in its reproductive habits, being the only cobra species to bear live young instead of laying eggs.

Although venom undoubtedly evolved as a means of immobilizing prey, many types of venomous snakes also use venom to deter potential attackers. Two groups of African cobras, and one Asian species, have evolved a particularly effective means of defense in this regard: spraying venom toward the eyes of an attacker. In "conventional" cobras the venom flows from a small aperture near the tip of the fang, but in the "spitters" this aperture is closer to the base of the fang and is rounded rather than elongate in

Anthony Bannister/NHPA

▲ The green mambas are the most arboreal members of the family Elapidae. Here an East African green mamba Dendroaspis angusticeps glides through a colony of weaver nests (Ploceus species), possibly in search of nestlings.

◄ By exhaling forcefully as venom drips from its fangs, a Mozambique spitting cobra Naja mossambica can spray venom at the eyes of an intruder up to 3 meters (9¾ feet) away.

▶ *If the bright warning coloration of the Arizona coral snake Micruroides euryxanthus euryxanthus fails to deter a predator, it defends itself by hiding its head, raising its tail, and everting its cloacal lining with a popping sound.*

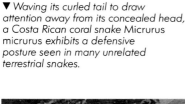

▼ *Waving its curled tail to draw attention away from its concealed head, a Costa Rican coral snake Micrurus micrurus exhibits a defensive posture seen in many unrelated terrestrial snakes.*

Michael and Patricia Fogden

shape. Also, the venom canal inside the fang reaches the outlet at a right angle to the tooth. Muscular action and vigorous exhalation cause two fine sprays that can travel several meters, and are aimed at reflective surfaces such as eyes. Venom delivered in this way will not kill a person, but can cause temporary blindness.

Asia contains many interesting species of elapids, including kraits and long-glanded coral snakes (in which the venom glands extend for the entire front third of the body), but the most famous Asian elapid undoubtedly is the king cobra. This magnificent animal is the largest venomous snake in the world, reaching more than 5 meters (16½ feet) in length. It is found in India and eastwards to southern China and the Philippines. The head of a king cobra can be as large as a man's hand, and these snakes are truly formidable when aroused. There are reliable reports of elephants dying within a few hours after being bitten by this species. Fortunately the king cobra is not an aggressive animal, even during the nesting season when the female deposits her eggs in twigs and foliage and remains to guard them. The king cobra feeds primarily on other snakes, including venomous species.

Coral snakes

The coral snakes belong to the subfamily Micrurinae. There are about 50 species (genera *Micrurus* and *Micruroides*) in the American tropics, mainly in South America. Most are brightly banded, and all are slender. Some species are aquatic, but most seem to be terrestrial foragers of the forest floor. Lizards and small snakes are probably their main prey, although some species also take mammals, birds, frogs, and invertebrates. The startling bands of color warn potential predators to stay away from these snakes, and some predatory birds have an innate fear of their color pattern. On other continents many semi-fossorial snakes are similarly marked with bright bands, even when the snake itself is harmless, so the bands may function to confuse predators encountering the snakes in dim light—as the snake thrashes around, the bands seem to flicker and fuse together and make it difficult for the predator to determine the exact position of the snake as it tries to escape. In the case of coral snakes, this coloration has been developed further as a warning symbol. Many unrelated harmless or mildly venomous snakes living in the same areas as the coral snakes have evolved color patterns that match those of the local coral snakes very closely, perhaps confusing predators and giving the snakes more chance to escape. Some coral snakes also employ an unusual posture to deflect the predator's attention from their vulnerable head: they hide their head among the coils of the body and the tail is waved around in the air in the way that you would expect the head to move. Any predator seizing the snake's tail by mistake is likely to receive an unpleasant surprise when the head suddenly appears.

Serpents of the sea

Two main evolutionary lines of fixed-front-fang snakes have taken to the oceans. The sea kraits, or laticaudine snakes (subfamily Laticaudinae) are brightly banded species of the Indo-Pacific region. Their common name comes from their resemblance to land kraits, a banded terrestrial Asian species of the genus *Bungarus*. Like the land krait, sea kraits have highly toxic venom but are extremely reluctant to bite, even in self-defense. One "sea krait" *Laticauda crockeri* is actually restricted to a landlocked lake in the Solomon Islands, but the other five species are truly marine. They feed on eels from the coral reefs but spend much of their time ashore on small coral islands.

Although they have flattened paddle-like tails that help them to swim rapidly, they retain broad belly scales and so are capable of moving around proficiently on land. They mate and lay their eggs ashore and can be found in large numbers on some small tropical islands.

The other group of fixed-front-fang snakes have become even more fully aquatic than the sea kraits, because they no longer need to return to land to breed. The "true" sea snakes or hydrophiines (subfamily Hydrophiinae), the latter name meaning "water-lovers", retain developing embryos within the female's body and produce live young, rather than laying eggs like the sea kraits. The tail is flattened laterally to act as an oar, the belly scales are reduced in size, and the nostrils are located on the top of the snout so that the snake can breathe even when most of its head is under water. The nostrils are sealed by valves that exclude water when the snake dives. The lung is much longer than in terrestrial snakes, extending almost the entire length of the snake, from just behind the head to the posterior end of the body cavity. It may play a role in adjusting the snake's buoyancy during diving, as well as storing oxygen-rich air for respiration. Some species can also take up oxygen directly from the surrounding water, through their skin, so that they can remain

◄▼ Laticaudines, like the yellow-lipped sea krait Laticauda colubrina (left), appear to have evolved their marine habits independently of the true sea snakes. They have retained their cylindrical shape and enlarged belly scales for crawling on land, which they frequently do to seek shelter, bask and mate. The pelagic sea snake Pelamis platurus (below), has a laterally compressed body for swimming and narrow belly scales, making movement on land extremely clumsy and difficult. A live-bearer, it does not leave the water, even to give birth.

► Named for a beak-like scale on the snout of mature males, the turtle-headed sea snake Emydocephalus annulatus is an Australian sea snake with a highly specialized diet of fish eggs.

submerged for long periods of time. Their adaptation to saltwater life includes specially modified glands at the base of the tongue, which function to concentrate and excrete excess salt from the snake's bloodstream. Most of the hydrophiids are so specialized for movement in water that they are almost helpless on land.

Biochemical evidence indicates that the hydrophiids might have evolved from the terrestrial Australian elapids, probably quite recently in evolutionary terms. They have been very successful, and the 50 species are widespread through the Indian and Pacific oceans. Most species are restricted to relatively shallow water and feed on fishes and their eggs around coral reefs. However, one species, the yellow-bellied sea snake Pelamis platurus, has adopted a pelagic lifestyle, drifting across the open oceans at the apparent mercy of the winds and currents. Its brightly spotted black and yellow tail serves as a warning to predatory fish not to attack the snake, which has very toxic venom. How can a snake drifting along on the surface of the ocean find and catch fish to eat? The yellow-bellied sea snake relies on the tendency of small fish to gather under any floating object, so the snake is soon "adopted" by a school of fish that swarm around its tail. But how can the snake seize the fish, when they are gathered around its tail and not its head? The answer is simple and elegant: the snake begins to swim backward, so that the fish now gather around its head. One rapid sideway strike, and the snake has its meal.

VIPERS

The highly venomous vipers tend to feed in a very different way from the fixed-front-fang snakes; they lie and wait to ambush unwary prey, especially mammals. Most vipers (and their close relatives, the pit vipers) are relatively heavy-bodied snakes, often beautifully camouflaged in their natural environments. They coil beside a mammal trail, or in the branches of a fruiting tree where birds are likely to gather, or beside a desert shrub where lizards will come for shade. There they wait for prey to wander within range. With their superb camouflage many of these snakes are almost invisible. Some actually lure prey within striking range by wriggling the tip of their tail, modified into an insect-like shape, to imitate the movements of a small invertebrate. Because they

do not have to move long distances actually searching for prey but must strike rapidly and accurately when a prey animal approaches, ambush-hunters tend to be relatively muscular and thus heavy-bodied. Because they eat large prey items, they need large heads. Because many of their prey items are not only large but also covered in fur or feathers, they need long fangs to penetrate deep into the victim's body and deliver the venom effectively. These snakes have thus evolved a very efficient means of killing large and potentially dangerous prey items. Unlike constricting snakes, which remain in contact with the prey while it struggles, vipers only need to inject venom with a quick strike and then wait. Even if the prey runs away to die, the snake can follow its scent trail.

Although the scientific name of the family, Viperidae, apparently comes from the Latin words *vivus* and *paro*, meaning "giving birth to live young", in fact there are many egg-laying (oviparous) vipers as well, both in the "true" vipers (subfamily Viperinae) and the pit vipers (subfamily Crotalinae).

"True" vipers

The 60 species of "true" vipers are widely distributed in Africa, Europe, and Asia. One species, the adder *Vipera berus*, has a remarkable geographic range from Britain, across Europe and Asia, to Sakhalin Island north of Japan. This species even lives in severely cold areas, extending into the Arctic Circle at 67°N and in

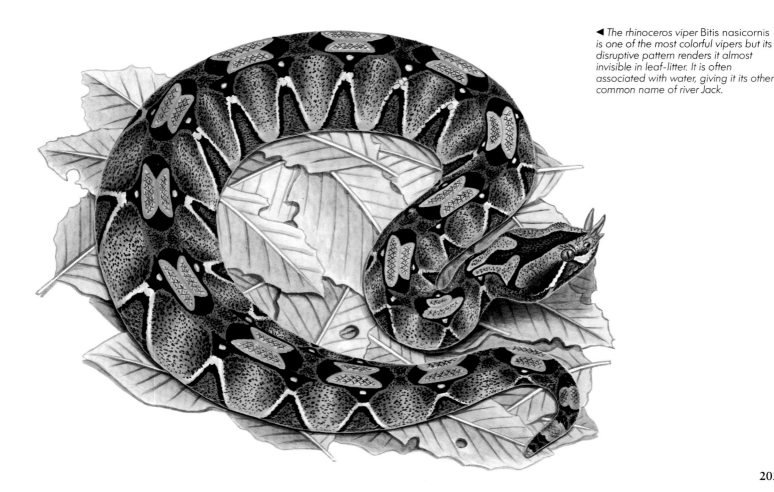

◀ The rhinoceros viper Bitis nasicornis is one of the most colorful vipers but its disruptive pattern renders it almost invisible in leaf-litter. It is often associated with water, giving it its other common name of river Jack.

parts of southern Siberia. In such cold areas, the adder can be active for only a few months each year and must spend the rest of the year deep within soil crevices to avoid the lethally low temperatures. The males emerge first in spring. After a few weeks of basking to ready themselves for reproduction, they shed their old skins and become far more brightly colored and active, roaming around in search of recently emerged females. If two males encounter each other near a female, they judge each other's size. The smaller male will usually flee, but if the two are evenly matched a battle may ensue. Instead of biting each other, they wrestle, with the two males intertwined until one is overpowered and gives up the fight. The winner then has the chance to court the female, and perhaps mate with her. Pregnancy takes about two months, and females spend most of their time basking so that the embryos develop at high temperatures and therefore more rapidly.

Females don't feed while they are pregnant, so they are often emaciated by the time they give birth. Many die at this time, and even those that survive are unable to gather enough energy in the autumn for reproduction the following year. Thus,

▼ A puff adder Bitis arietans rears its head in a threat posture. Normally undetected because of its cryptic coloration, this widespread African species puts on an impressive display when threatened, inflating its body and hissing loudly.

a female may be able to reproduce only once every two or three years, or even less often if food supplies are low or if the activity season is too short.

Vipers are found in tropical areas as well— for example, the large African puff adder *Bitis arietans*, Gaboon viper *B. gabonica*, and rhinoceros viper *B. nasicornis*. The saw-scaled viper *Echis carinatus* and its relatives are abundant in deserts of the Middle East. In these hot areas the reproductive cycles are likely to be quite different from those of the adder, but we know very little about tropical vipers in this regard. We do know that males of at least some species fight with each other, and that females of some species are egg-layers whereas others are live-bearers, but their detailed biology and behavior remains a mystery to researchers.

Pit vipers
The subfamily Crotalinae is the second major group of viperid snakes. Their common name, "pit vipers", comes from the deep pit between the eye and the nostril on each side of the head. Here there are sensory organs that detect heat. They are incredibly sensitive, detecting temperature differences of as little as 0.003°C. The ability to

recognize such tiny differences means that pit vipers can accurately locate and strike warm-blooded prey even on pitch-black nights. Most of the 140 or so species are ground-dwellers, but some are tree-dwellers and a few are semi-aquatic. They are widely distributed throughout the Americas, Europe, Asia, and Africa, but have not reached Australia or invaded the oceans.

▲ Beautifully camouflaged among fallen leaves, a West African gaboon viper Bitis gabonica rhinoceros awaits its next meal. This large species possesses the longest fangs of any snake, with reported lengths exceeding 5 centimeters (2 inches).

◄ This prey-eye view of an eyelash viper Bothriechis schlegeli shows clearly the heat-sensitive pits near the eyes, the chief diagnostic feature of the pit vipers.

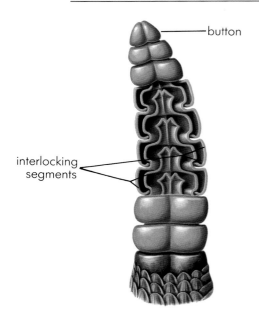

button

interlocking
segments

▲ Trademark of the rattlesnakes, the rattle is developed from enlarged and thickened scale covers that are retained after molting. These form interlocking segments that hit against each other when they move, producing the characteristic buzzing sound. A new segment is added after each molt, the oldest being the end segment, or button.

▼ The urutu Bothrops alternatus is one of a large group of closely related South American crotaline pit vipers, some of which are commonly referred to as fer-de-lance, a name originally given to the species found on the island of Martinique in the Caribbean.

The most famous of the pit vipers are probably the 30 species of rattlesnakes (genera *Crotalus* and *Sistrurus*) of North America, although three species (all of them belonging to the genus *Crotalus*) occur in Central and South America. These snakes have the tip of the tail modified into a remarkable warning device, the rattle. It is formed of specially shaped dry scales, one of which is added every time the snake sheds its skin (which doesn't happen on a regular annual basis, so the number of rattle segments cannot be used to determine the snake's age). When the tail is vibrated, the rattle segments move against each other to create the characteristic "buzz" that can often be heard from many meters away. It seems likely that the rattle evolved as a warning to large grazing mammals, such as bison, which share the prairie habitats of many rattlesnake species. Interestingly, one island population of rattlesnakes in the Gulf of California, living in an area without such hazards, has lost the rattle during its evolutionary history.

The other main group of pit vipers in North America, extending to Central America and Asia, is the genus *Agkistrodon* and its relatives. The North American representatives are the copperhead *A. contortrix* and the cottonmouth *A. piscivorus*. The cottonmouth (or water moccasin) derives its common name from the white coloration inside its mouth, exposed when the snake gapes in its dramatic defensive posture. Cottonmouths are swamp-dwellers and often

occur in large numbers in suitable habitats in the southern United States. On offshore islands they may live under heron rookeries, surviving almost entirely on fish dropped by clumsy herons.

The pit vipers have radiated extensively in Central and South America. A single egg-laying species, presumably the descendant of a separate pit viper invasion of the Americas, is the bushmaster *Lachesis muta* of lowland and lower montane rainforest. This huge snake, almost 4 meters (13 feet) in length, is a classic ambush predator, selecting a suitable ambush site beside a mammal trail and waiting, sometimes for weeks, until prey wanders within range. The scientific name *Lachesis* comes from one of the three Fates of Grecian mythology—the one who determines the length of the thread of life—because of the great danger to human life posed by a bite from this species.

Living in the same habitats as the bushmaster are a wide variety of pit vipers that give birth to live young, the most famous species being the fer-de-lance *Bothrops atrox*. In Central America this snake is also known by the Spanish name *barba amarilla* ("yellow beard") in reference to its yellow chin. Like the bushmaster it is a terrestrial ambush predator.

Pit vipers are diverse in Asia and include a range of terrestrial and arboreal forms. Some of the terrestrial species, like the habu *Trimeresurus flavoviridis*, are large brown snakes that strike readily when alarmed and cause many human

◄ Displaying a pattern unique among rattlesnakes, the lance-headed rattlesnake Crotalus polystictus is a small pit viper of the Mexican Plateau.

▼ Adopting the characteristic defense posture of rattlesnakes, a western diamondback rattlesnake Crotalus atrox vibrates its rattle to warn an intruder. Rattlesnakes probably evolved in North America when large grazing mammals like the bison more commonly posed a threat of trampling.

John Cancalosi/AUSCAPE International

▶ *Wagler's palm viper* Trimeresurus wagleri *(opposite) is a tree-dwelling pit viper of Southeast Asia. A tractable species, it is also known as the temple pit viper as it is frequently kept in "snake temples" where it is freely handled by the priests.*

fatalities in Japan every year. Others, like the palm viper *T. wagleri* found in the Sunda Islands and Malaysia, are spectacularly colored (bright green) tree-dwelling species that are very reluctant to bite humans; they are often kept as good-luck charms in "snake temples" and in trees near houses.

SNAKEBITE

Snake venom is a complex mixture of various chemical substances, mostly enzymes. It is a somewhat cloudy liquid, manufactured and stored in venom glands in the upper jaw and delivered to the prey either by injection through a hollow fang or by seeping into wounds caused by enlarged teeth. Some animals have very high resistance to particular snake venoms, whereas others are extremely sensitive to even small amounts of venom. The main constituents of snake venoms are enzymes such as proteinases which destroy tissue, hyaluronidase which increases tissue permeability (so that the venom can spread through the body more rapidly), phospholipases which attack cell membranes, and phosphatases which attack high-energy chemical compounds. The venoms of many colubrid snakes also contain L-amino acid oxidase, a substance that causes great tissue destruction. The venom of fixed-front-fang snakes (elapids and sea snakes) contains basic polypeptides to block nerve transmission and thus cause rapid death of the prey animal (or an unlucky human) by paralysing the diaphragm so that the victim stops breathing. In contrast, the venom of vipers has a high level of proteinase, which results in more severe tissue damage around the site of the bite and thus more profuse bleeding.

Which type of snake is the most deadly? The answer depends on several factors. The most potent venom is usually that of the fixed-front-fang snakes, the inland taipan *Oxyuranus microlepidotus* of central Australia being the

record-holder in this regard; a single bite from this species can contain enough venom to kill almost 250,000 mice. However, fixed-front-fang snakes have relatively short fangs and often do not produce as much venom as some of the large vipers. Also, some species are much more likely to bite than others, or are more likely to be encountered by humans.

Venomous snakes kill many thousands of people every year, especially in tropical or subtropical areas where people live in huge numbers and are often barefoot and where medical facilities are limited. For example, some authorities estimate that the saw-scaled vipers *Echis carinatus* in the Middle East may be responsible for about 20,000 fatalities each year. In Asia and Africa venomous snakes take a significant toll of human lives. In contrast, an average of fewer than five people per year die of snakebite in the whole of Australia, despite the fact that most of the Australian snakes are venomous, and some have the most toxic venoms in the world. Low population densities, adequate footwear, and excellent medical facilities are responsible for this low death toll.

CONSERVATION

Like most other living organisms, snakes have suffered at the hands of humans. The biggest threat to snakes, as to other animals, is the continued destruction of natural habitats. Rapid increases in human populations, and exploitation of the natural environment for logging, agriculture, and grazing, have decimated or eliminated snakes from many areas. The key to conserving snakes will be to conserve the places where they live. The snakes most at risk are those with a restricted distribution (like the Round Island boas) and those that depend on specific types of easily damaged habitats. For example, one small Australian elapid, the broad-headed snake

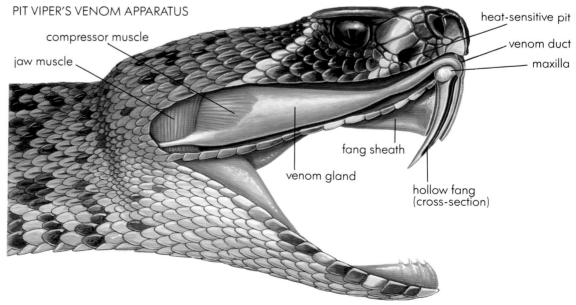

PIT VIPER'S VENOM APPARATUS

compressor muscle

jaw muscle

heat-sensitive pit

venom duct

maxilla

fang sheath

venom gland

hollow fang
(cross-section)

▶ *In most venomous snakes, venom is delivered by a highly evolved injection system, acting much like a syringe and hypodermic needle. The prey is first stabbed with the elongated fangs, then the venom glands are compressed by muscular contraction, forcing the venom through the venom ducts and the hollow fangs into the wound.*

Jean-Paul Ferrero/AUSCAPE International

DEFENSIVE STRATEGIES

Snakes defend themselves with a remarkable variety of behaviors. One of the most common, though least spectacular, is crypsis; the snake remains absolutely motionless despite the close approach of a potential predator. Viperid and boid snakes rely heavily on this behavior, and many of these snakes are beautifully marked with complex patterns that blend almost perfectly with their natural backgound. A large American copperhead coiled in sun-dappled foliage, or a puff adder in the leaf-litter, can be impossible to see unless it moves. Other snakes, especially more slender-bodied species, depend on speed for escape.

"Whipsnakes" from a variety of colubrid and elapid groups, although not closely related, share a common set of features such as large eyes, slender body, long tail, high selected body temperature, and a diurnal (day-active) life, which enable them to locate and capture fast-moving prey, usually lizards. These snakes move considerable distances and thus (unlike the "ambush" species) often encounter potential predators when they are in the open, far from cover. Camouflage is unlikely to be effective, and so they flee. These snakes are often unicolored or striped, unlike the "cryptic" species which more usually have a blotched pattern. Interestingly, this correlation between color pattern and mode of escape from predators even occurs within a single species, and sometimes within a single litter. Recent research on the garter snake *Thamnophis ordinoides* shows that among the newborn, blotched individuals are more likely to "freeze" and rely on camouflage when chased, whereas their striped litter-mates are more likely to rely on speed of escape.

If neither of these methods are effective, snakes still have many strategies to protect them from predators. Some hide their heads and wave their tails in the air, apparently to attract the predator's bites away from the head and neck to the less-vulnerable tail. The tail is often brightly colored in these species, and in pipe snakes is greatly flattened so that it resembles the hood of an angry cobra. A few American colubrids (*Gyalopion*) and coral snakes (*Micruroides*) take this even further by "cloacal popping", forcing air out of the cloaca to make a distinct popping noise. Why this noise should deter a predator is unclear. Nor do we know the significance of the bizarre defense display of the Australian bandy-bandy *Vermicella annulata*, which slowly raises body loops above the ground when threatened. Species such as the hognose snake (genus *Heterodon*) feign death by rolling onto their back, opening their mouth and extruding the tongue. Presumably, the predator loses interest in such an obviously "dead" snake.

Other snakes defend themselves more vigorously, striking at the attacker and often inflating part of the forebody to make themselves look larger and more formidable. Cobras (*Naja*) and hognose

snakes flatten their heads and/or necks horizontally, whereas Australian tree snakes (*Dendrelaphis*) and boomslangs inflate their necks dramatically. Sometimes brightly colored skin is visible between the expanded scales. Mambas and cottonmouths gape to display their fangs, while some large Australian elapids such as the king brown snake *Pseudechis australis* display chewing movements of their jaws.

The display can be auditory as well as visual. Most angry snakes hiss loudly in defense, and some, like the American bullsnake *Pituophis melanoleucus*, have a specially modified epiglottis to enhance the volume of the hiss. The rough-scaled desert viper *Echis carinatus* can produce a similar hissing sound by rubbing its body coils against each other. Many types of snakes twitch their tail-tips rapidly when alarmed, and this can produce considerable noise if the snake is lying in dry grass. The ultimate development of the tail as a sound-producing organ, however, is undoubtedly in the rattlesnakes. The "buzz" of a large diamondback *Crotalus atrox* is clearly audible for many meters, and is usually enough to convince a predator to look elsewhere for its meal.

Anthony Bannister/NHPA

▲ Buried to its eyes in sand, a Peringuey's desert adder Bitis peringueyi avoids detection by both predators and prey.

▼ A bandy bandy Vermicella annulata slowly raises loops of its body when threatened, possibly to confuse predators.

Pavel German

Hoplocephalus bungaroides, has become endangered because it relies on weathered sandstone boulders for shelter, the type of boulders that are rapidly being removed from natural habitats because they are popular as garden decorations. There is little point in "protecting" such a species with legislation, even international legislation, unless its habitat can somehow be preserved. This is a difficult task in many countries, where the immediate need to feed hungry people takes precedence over the needs of other species.

Some types of snakes are also threatened by commercial exploitation. The ones most at risk are large, brightly colored species whose skins are of value to the fashion industry. For example, pythons and boas are killed in many parts of their natural range because they are relatively slow-moving (easy to kill) and large enough to provide a valuable skin as well as a useful meal for local people. In some parts of Asia the killing of snakes is so intensive that the numbers of these animals have been considerably reduced. As a consequence, rats and mice previously kept in check by predatory snakes may become an agricultural problem. Some sea snakes are harvested for their skins and meat,

and huge numbers of sea kraits are taken every year from small coral islands where they come ashore to rest and reproduce. The ecological impact of this harvest has never been thoroughly investigated.

Many snakes are killed because humans hate and fear them for the supposed danger they represent. This fear is entirely legitimate in some cases, but in most areas snakes do far more good (by controlling agricultural pests) than harm. A high proportion of snakebites occur when people try to kill snakes, often in remote areas where the snake poses no threat to human safety. Most people cannot reliably distinguish between harmless snakes and venomous snakes—even in countries like the United States, where the two types are very different in appearance—and kill harmless species such as American watersnakes (genus *Nerodia*) in mistake for venomous snakes such as the cottonmouth in areas hundreds of kilometers away from where the venomous species actually lives. The infamous "rattlesnake roundups" of the southern United States represent one of the best publicized examples of the absurd enmity that many people feel for these magnificent but much maligned creatures.

RICHARD SHINE

▼ *Suffering the plight of most harmless snakes, the banded water snake* Nerodia fasciata fasciata *is commonly mistaken for a venomous species, in this case the cottonmouth* Agkistrodon piscivorus, *and killed on sight.*

AMPHISBAENIANS

Tanzania thread amphisbaenian
Chirindia rondoense
Body length: 9–12 cm (3½–4¾ in)
Diameter: 2–3 mm (⅛ in)

South American red worm-lizard
Amphisbaena alba
Body length: up to 72 cm (28½ in)
Diameter: up to 3 cm (1⅕ in)

CONSERVATION WATCH
■ Almost a third of the species
are known from only a single
specimen, so scientists have
no information about the size
of the population.

Worm-lizards, or amphisbaenians, have long represented an evolutionary
conundrum. Many specialists have at different times in their career reported
that they were lizards, only later (or earlier, in some cases) to consider them
as an independent offshoot within the order Squamata (to which lizards and snakes
belong). Certainly they are among the very few completely subterranean and self-
tunneling reptiles. They are so well matched to underground environments that they
seem very different from the lizards and snakes alive today.

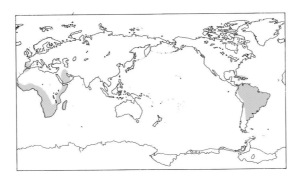

TRUE BURROWERS

All amphisbaenians seem capable of creating their
own tunnels. Burrowing is always done by the
head, and so the skull has been deformed during
their evolution to match the needs of whichever
burrowing method is used. The deformed skulls
also reflect the special needs of the feeding and
sensory systems of tunnel-dwelling animals. The
skulls are heavily ossified; and the small scales are
arranged in rings around the body. In all but one
species, the mouth is recessed into the lower
surface of the head.

Amphisbaenians generally have a long slender
body and a scaly skin (they shed the entire skin in
one piece); they have a transverse cloacal slit; and
the males have paired hemipenes, which are not
penises but rather pouches, one of which is
turned outward during copulation. These features
suggest that they belong to the order Squamata.
However, they differ from lizards and snakes in
several derived characteristics, such as a right-
rather than left-lung reduction, a uniquely shaped
egg tooth (used to pierce the eggs when hatching),
special skull bones, and the ring-like segments of
the skin. All living and fossil species discovered so
far have been classified in one of four families:

▼ *Amphisbaena fuliginosa of
South America is typical of the most
widespread family of amphisbaenians
or worm-lizards, the Amphisbaenidae.*

Chris Mattison

◄ Almost entirely subterranean in habits, worm-lizards are characterised by rather loose, baggy skins; blunt, heavily ossified skulls; underslung jaws; tiny eyes; and lack of evident ear-openings. They live in warm to tropical areas around the world. This is Bipes biporus of Mexico.

Trogonophidae (desert ringed lizards), Bipedidae (ajolotes), Rhineuridae, and Amphisbaenidae (worm-lizards).

Members of the family Bipedidae retain forelimbs, but external limbs are otherwise lost. Remnants of a shoulder girdle can be seen in the skeletons of the Trogonophidae, but not in the other families.

Most amphisbaenian species seem to be oviparous (egg-layers), but at least three cases of viviparity (giving birth to live young) have been documented.

THE SKIN

An amphisbaenian's head is always covered with shields larger than the scales of its body, and in some of the most advanced species these are fused together on the anterior part of the head. The scales of the trunk of the body are rectangular, or at least parallelograms, and are arranged on annuli (rings) that circle the body. Normally the segments on the underside are larger than those above. The cloacal slit is surrounded by a patch of enlarged segments, with the last annulus (ring) of the trunk sometimes having a row of precloacal pores, which in males may be involved during mating, although no observations as yet confirm this.

The skin of the trunk is loose around the circumference of the body, and muscles connect pairs of segments so that these can be narrowed or widened. There are three sets of muscles on each side of each segment: two reach backward and one reaches forward. These allow the animal to propel itself by rectilinear progression. A

portion of the body is fixed against the soil, while the backward-reaching muscles contract and pull up the rest of the body. Each forward-directed muscle then pulls the local portion of the skin out of contact with the soil and to a new, more forward position. It is quite different from the way snakes move; snakes fix the skin tightly near the middle of the back, whereas amphisbaenians have the skin free around the circumference.

Amphisbaenians are also capable of moving by concertina and lateral undulatory movements. Like many advanced snakes and limbless lizards, an amphisbaenian will use whatever movement is best suited to the soil and the shape of the tunnel—different propulsive patterns with different parts of their body.

The external appearance of different species gives us clues about the way they create their tunnels. For example, perhaps to reduce friction during penetration of the soil, adjacent segments of skin are fused not only on the head shields but also in the pectoral region of the spade-snouted species, for this part of the body pulls forward, rubbing against the ground when the head is lifted to compress the soil.

PREY AND PREDATORS

Worm-lizards are effective predators, and they can bite pieces out of larger animals. Even the smaller species have massive jaw-closing muscles. The upper jaw has five to nine teeth on each side, with a single central tooth that fits between the enlarged teeth of the lower jaw. Amphisbaenians can therefore exert powerful bites, and their

Carl Gans

▶ A white-bellied worm-lizard *Amphisbaena alba* from Brazil gapes in threat. Worm-lizards are formidable predators, with sharp teeth and powerful jaws. Members of the genus *Amphisbaena* are locally (but erroneously) considered poisonous, and are called "ant-kings" from the widespread belief that they are raised by ants or termites. Recent observations confirm the connection with these insects, but the exact relationship remains a mystery.

Martin Wendler/NHPA

HEARING UNDERGROUND

Terrestrial reptiles generally have external eardrums on the sides of the head or at the base of an external ear canal. From these a stapedial link crosses the air-filled middle ear to an oval window that opens into the liquid-filled tube of the inner ear. Here, rows of hair cells transduce the sound vibrations into electrical signals which pass to the brain, where they are perceived as sounds. As signals reach the two ears at slightly different times and at different magnitudes (the head serving as a sound shadow), the animal not only hears the sound but may be able to orient to it.

Reptiles living underground tend to lose the external eardrums and sometimes even their middle ears. This suggests not only that their ear openings are protected from predator or parasite attack, but also that they rely more on other senses. Members of three of the four families of amphisbaenians, however, have developed a unique hearing system that lets them sample sounds emanating from the tunnel ahead of them. Bones of the throat now connect with the stapedial plate on either side of the face, and scales of the skin on the face act as a substitute forward-facing eardrum. Experiments have shown that the system indeed detects airborne sounds,

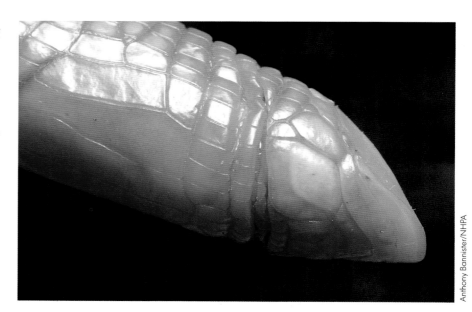

Anthony Bannister/NHPA

and is as sensitive to these vibrations as to those applied directly to the skull. This indicates that they hear and orient to the sounds of prey, and possibly to others of their own kind.

▲ Specialized scales on the face function as substitute eardrums in an auditory system unique to worm-lizards.

Chris Mattison

interlocking teeth can cut and tear out pieces from larger prey; worm-lizards have been observed in their tunnels biting and then spinning their body to free the portion of flesh.

Most reports suggest that amphisbaenians are omnivorous, attacking and subduing any small animals encountered in their environment; but many of these are observations on captive animals, and we still know relatively little about their feeding habits. Seasonal changes of diet may well be important. Recently we learned that some centuries-old reports of the association of the giant *Amphisbaena alba* with the nests of South American leafcutter ants are indeed correct, although we don't yet know what may be the mutual benefit to the worm-lizards and the ants.

Amphisbaenians are in turn threatened by many predators. There are those that forage on the surface, such as birds and small mammals; indeed, one seemingly rare worm-lizard was first described by a scientist dissecting the crop of a small eagle. They are also in danger from pigs that dig in the soil and root out any "protein" encountered, and monkey troops that forage over wide regions, lifting pieces of wood and rocks under which amphisbaenians may shelter. Finally, there are tunnel predators, such as some African snakes, each of which is claimed to specialize on a particular local amphisbaenian.

Perhaps 30 percent of amphisbaenians show a darkened color pattern suggesting camouflage,

which suggests that they may make occasional visits to the surface. A few species have been observed basking on the surface. And when the skin is darkly pigmented it coincidentally seems to be less permeable to water than the skin of deeply burrowing species, which normally has substantial water loss. Perhaps some of them do make voluntary visits to the surface. Alternatively they may move to the surface in response to flooding after heavy rains and to attacks by various predatory ants.

All but the African amphisbaenians have short tails, the end of which is commonly reinforced with dense connective tissues that may be supported by fused terminal vertebrae. In spite of this, many species show "caudal autotomy", or voluntary tail-shedding—they break off their tail when a predator gets hold of it. The broken tip may move and distract predators, whereas the stump may heal, so perfectly that some people have been fooled into thinking that the individual belongs to a "new", "short-tailed" species. Other species use their tail differently for defense: the tip has a cone-covered surface which attracts dirt and thus supplies a block against any predator following the worm-lizard down its tunnel.

THE AMPHISBAENIAN FAMILIES
The family Trogonophidae is the most distinct group of amphisbaenians. Species occur in North Africa and also from western Iran across the

▲ *Except for their obvious scales, amphisbaenians might well be mistaken for caecilians in their size, shape, behavior and general appearance. A scarred tail stump on this* Amphisbaena fuliginosa *confirms that worm-lizards, like many skinks and lizards, exhibit caudal autotomy — that is, they can voluntarily shed their tail if grabbed by a predator.*

Arabian Peninsula to Aden, Socotra Island, and eastern Somalia. They tunnel using an oscillating movement, and their body structure has special modifications for this such as a downward-pointing tail, which the animal digs in to give extra force for burrowing. The teeth are permanently fused to the crest of the bones, rather than being in a groove and attached to the sides of the bones. The species that live in Arabia and Somalia show the most advanced characteristics and have a general tendency to become shorter and stouter.

The family Bipedidae consists of three species (genus *Bipes*) in northwestern Mexico. All of them have hands, which lie so far forward and are so large that the animals bear the popular name of "little lizard with big ears". The fingers are used during tunnel-building to scrape away soil and widen the tunnel.

Florida is the home of the only remnant species of the once-widespread family Rhineuridae. Fossils have been found across the western United States, dating back in time to the Paleocene, about 60 million years ago. *Rhineura*

floridana has a spade-shaped head, but with a different arrangement of bones from those in the other amphisbaenian families.

The true amphisbaenids (family Amphisbaenidae) are more than 120 very diverse species with a variety of tunnel-forming specializations. The most primitive occur in the West Indies and northern South America. Other species range south to Patagonia and cross the Andes in Peru. In Africa, they live south of the Sahara Desert, but not in the Central African highlands or at the southernmost tip of the continent. Some have a keel or crest on the head, others have a spade-shaped head.

Species in the genus *Blanus* may represent a fifth family. They live in Spain, North Africa, Lebanon, northern Israel, and northern Iraq. Fossils have been found in north-central Europe. Some of the species alive today have several primitive features such as a mouth that is not recessed under the snout and residual hind limbs, but we really need to know more about them before classifying them in one family or another.

CARL GANS

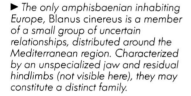

▶ *The only amphisbaenian inhabiting Europe, Blanus cinereus is a member of a small group of uncertain relationships, distributed around the Mediterranean region. Characterized by an unspecialized jaw and residual hindlimbs (not visible here), they may constitute a distinct family.*

Vial/Jacana/AUSCAPE International

◄ Known locally by a name that translates as "little lizard with big ears" from the size and far-forward position of its forelimbs, Bipes biporus and two other species together constitute the family Bipedidae, restricted to northwestern Mexico. All other worm-lizards are limbless.

BURROWING STYLES

An amphisbaenian makes a tunnel by forcing its head into the soil using either compression of the soil in front of it, or by oscillation, in which the end surface of the tunnel is regularly scraped away, later to be forced into the walls. Most species use a compression method (perhaps one of several variants described below) to make permanent tunnels in many different types of soils. Oscillation methods of tunnel-building are used by members of the Trogonophidae family; they mostly live in sandy soils, which are less compressible than other soils, but the grains may be packed in different patterns.

Soil-compressing amphisbaenians force their heads directly into the soil, and some use slight sideways movements to enhance penetration. In the simplest pattern, the amphisbaenian's snout is somewhat rounded, and a series of pushes drives it forward. Between the drives, some species pull up the back of the body, curve the neck, and then produce the next penetrating drive by straightening these curves.

Intermittent penetration by soil-compressing species—who drive the head in part-way then rotate it, thus compressing the loose soil into the tunnel wall—has resulted in two quite different modifications to the amphisbaenians' skeletons. In the first, the skull has become spade-shaped; the head drives into the tunnel end at a fairly low level, and is then rotated upward, the blade of the spade compacting the loose soil into the roof of the tunnel. In the second, the skull has formed a spectacular crest, or keel, and the amphisbaenian widens its tunnel by packing the soil into either side. In both types the rotation of the skull involves the axial muscles of the segments behind the head, so tunneling in hard soils is possible without an increase in the animal's body diameter, as the force-generating muscles are distributed along the trunk. The spade-snouted animals do

have an advantage in that they drive the head upward only and thus need only a single set of axial muscles. In contrast, the keel-heads tend to alternate compression to left and right, and only half of the available musculature is involved in either drive.

The oscillation method of tunnel-making may require a continuous application of force against the end of the tunnel where the surface is being scraped. Some species have a keratinized edge on each side of the head, which serves as a scraper. The scraping movement is complex, as the head twists about two different axes. A short and stout body helps the amphisbaenian to apply force from its tail; a pointed tail can be dug in easily to apply this force. A body shape that is triangular or rectangular in cross-section prevents the amphisbaenian from spinning during oscillation.

▲ Agamodon angeliceps is a member of the family Trogonophidae, a group that tends to live in sandy soils and uses oscillating head movements in tunneling. The somewhat rectangular cross-section of the body resists twisting, while the rim on each side of the spade-shaped face-plate shaves soil toward the roof, from one side, then the other.

TUATARA

D espite appearances, tuatara are not lizards. They are the sole living members of the order Rhynchocephalia ("beak-headed"). Fossil rhynchocephalians were small to medium-sized reptiles that were common throughout the world between about 225 and 120 million years ago, long before the first dinosaurs appeared. Later their numbers declined, and about 60 million years ago they apparently became extinct everywhere except in New Zealand.

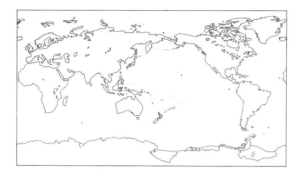

DISTINGUISHING FEATURES

Tuatara are of extraordinary zoological interest and are often referred to as "living fossils". However, recent research suggests that tuatara display many advanced features when compared with their nearest extinct relatives. They cannot, as is often claimed, have persisted unchanged from their ancestors for 100 to 200 million years.

Tuatara differ from lizards in many ways. For example, they lack external ears; they have hook-like extensions on some of their ribs (a bird-like feature); they have two large apertures on each side of the skull immediately behind and above the cavity housing the eyeball (lizards have only one); and the males do not have a penis.

Tuatara have a legendary "third eye", which is part of a complex organ situated on the top of the brain. This has a lens, retina, and nerve connection to the brain, but early in the growth of the tuatara the organ is covered by opaque scales. Many lizards also have a "third eye" of similar complexity. While in lizards this organ is involved in regulation of body temperature, experimental tests have failed to reveal the function, if any, of the "third eye" in tuatara.

Tuatara may be gray, olive, or occasionally brickish red in color, and are spotted to varying degrees. In the nineteenth century several species and subspecies were named on the basis of differences of color and body proportion, but most differences are now known to occur within populations. Thus the tuatara became almost

▶ *Tuatara were once widespread throughout New Zealand, but were exterminated on the main islands within the last few centuries, apparently through a combination of hunting, changes in land use, and introduced predators such as rats. Their populations are now restricted to about 30 small offshore islands, where, even on many of these last refuges, their status remains precarious.*

Michael Schneider/New Zealand Geographic

John Cancolosi/AUSCAPE International

universally regarded as a single species, *Sphenodon punctatus*. However, work still in progress on genetic variation among tuatara populations has supported the early scientists' contention that at least two species of tuatara exist. The population living on North Brother Island in Cook Strait, is now considered by researchers genetically different enough to be regarded as a distinct species, *Sphenodon guntheri*.

Male and female tuatara can be easily distinguished. The male is substantially larger than the female and is generally twice as heavy. The adult male has a striking crest along the back of the neck and another down the middle of the back, both of which bear conspicuous soft spines. The female also has neck and body crests, but neither crests nor spines are well developed. The

male has a proportionately narrower abdomen and a larger head than the female.

Like most lizards, tuatara are capable of caudal autotomy—shedding their tail as a means of escaping from enemies. The lost portion regrows but differs in color and pattern from the original and is shorter.

WHERE TUATARA LIVE
Tuatara or their ancestors may have reached New Zealand over 140 million years ago, when it was part of the Gondwana landmass, but no fossils older than 25,000 years have been found. Those located show that tuatara once ranged throughout the two main islands of New Zealand. Now, however, they are restricted to 30 small islands off the northeast coast of the North Island and in

▲ *In tuatara the sexes differ considerably in size. A large male may extend 60 centimeters (2 feet) from snout to tail-tip, and weigh over 1 kilogram (2 pounds), but females seldom exceed 500 grams (17½ ounces) in weight. Males also usually have more conspicuous spines on the nape and back.*

Cook Strait. Half of these islands have an area of 10 hectares (25 acres) or less.

Today, there are probably more than 100,000 tuatara, but *Sphenodon guntheri* occurs naturally on only one small, 4 hectare (10 acre) island, which supports fewer than 300 adults. Stephens Island in Cook Strait supports at least 30,000 *Sphenodon*

punctatus and possibly more than 50,000. There, exceptional density equivalents of over 2,000 tuatara per hectare (2½ acres) have been recorded.

Most tuatara islands are cliff-bound and difficult to access. They are often exposed to strong winds and the natural vegetation is stunted. Many have been cleared by Maori or European people for

AND THEN THERE WERE TWO . . .

Alison Cree

In the nineteenth century two living and one extinct species of tuatara were named. Later it was contended that differences in features such as color and body proportion, upon which the proposed species were distinguished, were no greater than could be observed within individual colonies. Thus the tuatara became universally regarded as a single species, *Sphenodon punctatus*.

During the southern summer of 1988–89 zoologists from Victoria University of Wellington, New Zealand, began a comprehensive survey of the genetic and morphological variation in tuatara. Expeditions have been made to 24 of the 30 islands on which tuatara populations are thought to survive. Six islands have not been visited because of the extreme rarity of tuatara or

difficulties of access. From blood samples taken from animals, three distinct groups of tuatara have been identified. Most importantly, the population living on North Brother Island in Cook Strait, was distinguished as a separate species, *S. guntheri*, a name first proposed in 1877. Among other populations, a northern group and a western Cook Strait group were distinguished. Further, analysis of seven body characters (for example, body length, length of longest neck spine) showed that the same three groups could be differentiated morphologically, although considerable overlap occurred. Western Cook Strait and northern populations were thought sufficiently distinctive from each other to warrant assigning them subspecific status within *S. punctatus*.

▲ *An adult male of the North Brother Island tuatara population, Sphenodon guntheri. Tuatara have long been treated as a single species, but recent research has demonstrated the existence of two species, very similar in appearance but differing substantially in genetic structure.*

cultivation and grazing. Lighthouse stations (now automated) exist on some. The major factor common to all is the presence of breeding populations of burrow-nesting seabirds (petrels and shearwaters) which, in part through the influence they exert over soils and vegetation, influence the tuatara.

Many tuatara islands support several species of petrels and shearwaters, but islands where tuatara still maintain their numbers are rodent-free and have very large breeding populations of the fairy prion *Pachyptila turtur* and/or the diving petrel *Pelecanoides urinatrix*. These small petrels provide tuatara with many benefits, principally housing and food. They excavate burrows in which many tuatara live (although the reptiles also dig their own), and by turning over the soil and incorporating their mineral-rich guano they create conditions that may increase production of ground-dwelling invertebrates which form the bulk of the tuatara diet. Tuatara also eat lizards, the adults, chicks, and eggs of small petrels, and even their own young.

FEEDING AND BREEDING

Tuatara are active mostly at night and spend the day in burrows or, if it is sunny, basking at burrow entrances. Only one tuatara at a time has been found in each burrow. As with reptiles generally, the number of tuatara abroad at night is largely dependent upon environmental temperature. Maximum activity occurs between 17° and 20°C (63° and 68°F)—low for a reptile—then declines steadily as temperature falls, becoming negligible below 7°C (45°F). Other weather variables can also affect activity. For instance, tuatara slowly dehydrate unless they have access to water. There are no ponds or streams on many islands where they live, so after prolonged dry spells the reptiles are particularly active on rainy nights, when they may frequently be found soaking in puddles.

When hunting, tuatara usually adopt a sit-and-wait strategy. Prey moving in their vicinity is seized—smaller items with the tongue, and larger items by being impaled on the incisor-like teeth. Small, hard items such as beetles are crushed and dismembered, and larger prey are literally carved up. On each side of the mouth the tuatara has two rows of teeth on the upper jaw and one row on the lower jaw; the lower row fits between the two upper rows, and when chewing, the lower jaw is moved back and forth in a sawing motion, akin to the action of a carving knife.

Courtship and mating take place in late summer–autumn (January–March). During January males establish territories, each including the territories of several females. They display at conspicuous locations within their territories, presumably to attract females and to advertise their presence to other males. This involves

Michael Schneider/New Zealand Geographic

▲ *Tuatara are essentially nocturnal, but often sunbathe outside their burrows by day.*

increasing the size of the trunk and throat by inflating the body and elevating the crest and the spines.

Territorial encounters between males take the form of "face-off" displays and fights. In face-off displays, two males line up side by side about a meter (3 feet) apart, often with their heads facing in opposite directions, then slowly gape open their mouths and rapidly snap them shut. One male may be chased away, or a fight may ensue. During fights, one male locks his jaws around the head or neck of the other male, and the two scuffle on the ground, often emitting croaking calls at the same time. Facial wounds, broken jaws, and lost tails are not uncommon.

If a female is attracted, a displaying male will circle her using a slow, intermittent, stiff-legged walk. This may continue for about 20 minutes and ends when she either disappears down a burrow or allows the male to mount her. The male mounts the female from behind and moves to lie on top of her flattened body, using his hind legs to lift her tail so that his vent can come into contact with hers for transfer of sperm.

Eggs are laid in the following spring or early summer (October–December). Females gather at sunny, unforested sites to nest (the soil in forests being too cold for successful incubation). A female may first dig chambers or blind tunnels as trial nests, but eventually lays her clutch of up to

19 soft-shelled eggs (generally seven to ten) in one nest during a single night. Several more nights are spent filling the nest entrance with soil. She may return to her nest for eight consecutive nights, even remaining alongside it by day. This is because other females often dig up nests already containing eggs, then lay their own eggs at the same site. When at their nests, females may defend them vigorously from other females, such encounters being every bit as intense as those between territorial males. On average, females produce eggs just once every four years; males have an annual reproductive cycle.

Tuatara have one of the longest incubation periods of any reptile—12 to 15 months, the actual time depending on temperatures during this period. Temperature can also affect the sex of the hatchlings. Unlike eggs of most reptiles, those of the tuatara temporarily cease development during winter, when conditions can be quite cool.

As in crocodilians and turtles, young tuatara are born with a horny "tooth" on the tip of their snout, which they use to cut their way out of the parchment-like egg shell. Upon emergence they average 5 grams (¼ ounce) in weight and 100 millimeters (4 inches) in length. They retain a remnant of the allantoic sac, which shrivels and falls off within a few days; the horny tooth falls off after about two weeks. Young are normally a cryptic fawn and brown color. They can move very quickly and are not often seen. They are

▼ Mating tuatara Sphenodon punctatus. They mate annually, in February and March, but females lay eggs only every four years or so. Unlike most reptiles, the male tuatara has no penis, and copulation depends on direct contact between the partners' vents. No pair bond is formed; the two remain together for about an hour, then go their separate ways.

PRESERVATION OF A "LIVING FOSSIL"

Tuatara live only in New Zealand. They were once widespread over the two main islands, but during the past 150 years have become extinct there, as well as on at least 10 offshore islands. Habitat destruction and predation by cats, rats, pigs, and other mammals introduced by Maori and European settlers are probably the main causes of extinction.

While about 30 populations are thought to survive, eight are vulnerable or endangered because of the presence of an introduced rat (the kiore, *Rattus exulans*). During recent surveys, no juvenile tuatara were found on seven of the eight rat-inhabited islands. Although we don't yet have experimental evidence, kiore probably prey on tuatara eggs and juveniles, and compete for food with juveniles and adults.

Ten years ago eradication of introduced rodents from offshore islands seemed an impossibility, but recent success with islands of up to 2,000 hectares (5,000 acres) has changed that view. In a typical rodent eradication program, baits containing an anticoagulant poison are placed systematically over an island. Poisoning is carried out in spring, when the number of rats is low and there is little food available. Monitoring continues for at least a year after the poisoning before the island is considered rat-free. Tuatara are removed before poison is laid, and any progeny raised in captivity can be released on the island once it has been made free of rats. By 1998 rats had been removed from seven of the eight rodent-infested islands and tuatara re-introduced to two of them.

Few attempts to breed tuatara in captivity have been successful, although recently, significant advances have been made. Enhanced knowledge of the extensive behavioral repertoire of tuatara suggests that social interactions should be

Brett Robertson

encouraged in captivity (previously adult tuatara were often maintained in single pairs). We now know how to incubate eggs artificially, and that eggs can be induced from females ready to lay by administering a hormone called oxytocin. Using these techniques, eggs have been collected from wild females of the North Brother Island tuatara. This species occurs on only one small, 4-hectare (10-acre) island, which supports fewer than 300 adults. The eggs were successfully incubated in the laboratory, and the hatchlings used to establish a second population of this vulnerable species.

▲ There is no parental care in tuatara. Females lay a clutch of (usually) 7 to 10 eggs in underground nests that are then abandoned. After 12 to 15 months of incubation, the young tuatara hatches, using a special egg-tooth on the snout, visible on this youngster, Sphenodon guntheri, to tear open the eggshell. It then burrows its way to the surface. Tuatara eggs have been incubated successfully in research laboratories.

most active during the day—possibly a strategy made necessary by the cannibalistic tendencies of the larger adults.

As with the duration of incubation, the growth rate of tuatara largely depends on temperature. The warmer the conditions, the faster the growth. At the southern part of their range animals mature at between 11 and 13 years, and to the north about two years earlier. Sexually mature tuatara may continue to grow for a further 20 years or more; on Stephens Island most animals reach full size at somewhere between 25 and 35 years of age. A number of Stephens Island adults recaptured after 30 years showed no appreciable growth, so the life-span of the species must be at least 60 years. One animal, collected as an adult, is known to have survived in captivity for more than 70 years.

DONALD G. NEWMAN

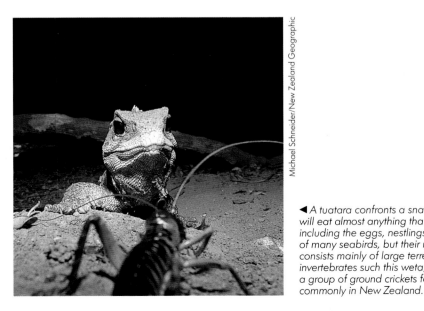

Michael Schneider/New Zealand Geographic

◄ A tuatara confronts a snack. Tuatara will eat almost anything that moves, including the eggs, nestlings, and adults of many seabirds, but their normal diet consists mainly of large terrestrial invertebrates such this weta, one of a group of ground crickets found commonly in New Zealand.

CROCODILES & ALLIGATORS

The order Crocodilia includes the largest of the living reptiles, and they are among the largest of the vertebrates that still venture onto land. During the Mesozoic era 245 to 65 million years ago, the Archosauria ("ruling reptiles") dominated the land, but the only giant archosaurs that have survived to modern times are the crocodilians. The dinosaurs left other descendants—the birds—but it is only the crocodilians (crocodiles, alligators, caimans, and gharials) and a few turtles that today reflect the majesty of the time when giant reptiles ruled the Earth.

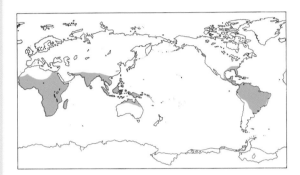

PREDATORS IN TROPICAL RIVERS

Most of the 23 or so species of crocodilians occur in tropical parts of the world, but a few, notably the American and Chinese alligators, extend into temperate regions. All species are semi-aquatic and do not venture far from estuaries, swamps, lakes, streams, or rivers. Their generally large size and short legs mean that they are agile on land only over very short distances. In the water, however, they are superbly adapted predators.

The external characteristics that differentiate crocodilians from other reptiles are mainly adaptations to a semi-aquatic lifestyle. For example, the elongate head and numerous peg-like teeth are characteristic of many aquatic animals with backbones, from fish to dolphins. Such heads are probably too ungainly for purely land-dwelling creatures. The eyes, with their transparent protective third eyelids (nictitating membrane), and the nostrils, with their watertight valves, are set high on the head so that these remain at the surface while the rest of the animal is completely submerged. Most swimming is done by strokes of the powerful blade-like tail.

On land, crocodilians can lift their bodies high off the ground and walk with their legs positioned almost directly below the body, as does a mammal. However, they cannot maintain this position at speed, and most species revert to a typically reptilian splayed-leg belly slide if hard-pressed. One species, Johnston's crocodile *Crocodylus johnsoni*, gallops over short distances to

escape back to the water, sometimes with all four limbs off the ground at the same time.

Internally, crocodilians have some features that are more similar to those of mammals than those of other reptiles. Crocodilians have a bony secondary palate. This, combined with a fleshy valve at the base of the tongue, completely isolates the respiratory system from the mouth, allowing them to open their mouths underwater without risk of drowning. The intestinal cavity is separated from the pectoral cavity by a muscular septum, analogous to the diaphragm of mammals, presumably increasing the efficiency of ventilation of the lungs. The heart has four separate chambers, but crocodilians have retained an

aperture, the foramen of Panizza, between the left and right ventricles and are able to shunt blood to the regions of the body that most require it while warming in the sun or diving.

Three families (or subfamilies) of living crocodilians are recognized by taxonomists. Although their lineages have been distinct for tens of millions of years, differences between the families are few.

THE ALLIGATOR FAMILY

Alligators and caimans are classified in the family Alligatoridae and are therefore known collectively as alligatorids. In all alligatorids the teeth of the lower jaw fit into pits in the upper jaw and cannot

▼ *Lord of the waterways: the Nile crocodile* Crocodylus niloticus. *Although agile only for short distances on land, there are few more dangerous or formidable predators than the crocodile, even to creatures much bigger than humans. Adult male Nile crocodiles sometimes exceed 1 tonne (1 ton) in weight and 5 meters (over 16 feet) in length when fully grown.*

▲ *Restricted to the southeastern United States, the American alligator* Alligator mississippiensis *is one of only a few crocodilians whose range extends well beyond the tropics. Its blunt, broad jaws help it catch prey in thick vegetation.*

be seen when the jaws are closed. The scales on the underside of the body have no sensory pits.

The two species of true alligators occur in widely separated regions of the world. The American alligator *Alligator mississippiensis*, which grows to 6 meters (20 feet) in length, occurs in the southeastern United States. The Chinese alligator *A. sinensis* inhabits the lower Yangtze River and its tributaries in China; it rarely exceeds 2 meters (6½ feet). Both species have broad heavy heads, which are presumably adaptations to living in heavily vegetated swamps. In open air or in water, thin snouts would be more maneuverable, but a heavy head has more momentum to catch prey by smashing through intervening vegetation. When the ponds and swamps inhabited by the American alligator freeze over, as they do occasionally, most of the larger animals survive by lying in shallow water and maintaining a breathing hole through the ice. Occasionally their snouts are frozen into the ice, but provided their nostrils are not covered, they usually survive. Most other species of crocodilians would be killed by such low temperatures.

Alligatorids in South and Central America are all called caimans. The largest is the black caiman *Melanosuchus niger*, which can grow to more than 6 meters (20 feet). Biochemical evidence indicates that it is more closely related to other caimans

than to alligators, but externally it is similar to the American alligator. Unlike most crocodilians, which may be vividly marked as hatchlings but soon assume the drab colors of adults, the black caiman remains distinctively marked throughout life. Hatchlings have light gray heads and black bodies patterned with lines of white dots. As they grow, the gray on the head becomes brown and the lines of white dots fade, but even adults longer than 5 meters (16½ feet) are more colorful than the hatchlings of most other species. Black caimans inhabit flooded forests and grassy lakes around slow-flowing rivers throughout the Amazon basin in Brazil and to the north in the coastal rivers of the Brazilian state of Amapa, in French Guiana and in Guyana. They were once the most commonly seen caimans throughout their range, but hunting (for their skins) has drastically reduced their numbers. Part of the problem seems to be that the black caiman occurs in the same areas as the common caiman, and that this species sustains hide hunting in areas long after hunting for the black caiman alone would have become uneconomic.

The genus *Caiman* has two species: the common caiman *C. crocodilus*, which occurs from southern Mexico to northern Argentina; and the broad-snouted caiman *C. latirostris*, which occurs in streams and marshes along the coast of Brazil from the state of Rio Grande do Norte south to Uruguay, as well as in the inland basins of the São Francisco, Doce, Paraíba, Paraná, and Paraguay rivers in Brazil, Paraguay, and Argentina. Both species are extremely adaptable and can be found in a variety of habitats, including large rivers in tropical rainforest, seasonally flooded savannas, permanent swamps, and mangroves. The common caiman is by far the most hunted crocodilian in the world, accounting for 60 to 80 percent of the skins in trade. However, its bony skin is of comparatively low value, and a hide of this species typically attains only about a tenth of the price paid for *Alligator* or *Crocodylus* skins. Dwarf caimans of the genus *Paleosuchus* are among the least studied of the crocodilians. Schneider's dwarf caiman *P. trigonatus* occurs throughout the forested areas of the Amazon and Orinoco basins and the coastal rivers between them; it is rarely found away from thick tropical rainforest. Cuvier's dwarf caiman *P. palpebrosus* occurs throughout the same region but generally lives in more open habitats such as gallery forest in savannas, or the margins of large lakes and rivers. It also extends across the high plains of the Brazilian shield, south to northern Paraguay.

Dwarf caimans are sometimes called smooth-fronted caimans because they lack the bony ridge between the eyes that is typical of *Caiman* and *Melanosuchus*. Their skins are heavily ossified—a build-up of calcium carbonate and calcium phosphate in the connective tissue—and are much less flexible than those of most other crocodilians. Even their eyelids are heavily ossified, and they have been compared to turtles because of this dermal armor. Whereas Schneider's dwarf caiman has a fairly typical head shape, the skull of Cuvier's dwarf caiman is high, smooth, and dog-like. There are as yet no convincing explanations as to why it should have

Michael and Patricia Fogden

evolved this unique skull shape. Both species are small. The total length of Schneider's dwarf caiman is about 1.7 meters (5½ feet) for an adult male and 1.3 meters (4¼ feet) for a female. Cuvier's dwarf caiman is probably the world's smallest crocodilian, at about 1.5 meters (5 feet) for an adult male and 1.2 meters (4 feet) for a female. Both species have rich brown eyes, unlike the yellowish eyes of most other crocodilians.

THE CROCODILE FAMILY
Members of the family Crocodylidae have a notch in the upper jaw on each side which accommodates the fourth tooth in the lower jaw when the mouth is closed, this tooth is therefore

▲ *Lacking extensive osteoderms (bony buttons) in the belly scales, the black caiman* Melanosuchus niger *of Amazonia yields a so-called "classic" skin, much in demand for the manufacture of luxury items like handbags. Intense hunting during the 1950s and 1960s quickly reduced the species to critically low population levels.*

▼ *Cuvier's dwarf caiman* Paleosuchus palpebrosus *is the smallest and most heavily armored crocodilian, with even its eyelids protected by bony plates.*

Stan Osolinski/Oxford Scientific Films

▲ A group of Nile crocodiles basking on a river bank in Kenya.

Crocodylids possess salt glands on the tongue, which can be used to excrete excess salt in saline environments. Even species that live in totally fresh water have functional salt glands, and it is thought that the tolerance of crocodylids for salt water may explain their wide distribution throughout the tropics.

The genus *Crocodylus* includes 13 fairly similar species. The last was named in 1990, and in the near future another species of the Indo-Pacific region will probably be reclassified as two separate species. All are moderate to very large, with narrow snouts. None has a broad snout like an alligator or a snout as thin as that of a gharial.

The Indo-Pacific crocodile *C. porosus* ranges from India to northern Australia and the Solomon Islands. The presence of this very large species in estuaries and mouths of large rivers may serve to prevent regular contact between the six or more freshwater species that occur on the major land-masses in the region. This species is probably the largest of the crocodilians, and lengths of more than 7 meters (23 feet) and weights of more than 1,000 kilograms (almost 1 ton) have been recorded.

In the New World the American crocodile *C. acutus* seems to have a similar role; it occurs in estuaries from the southern tip of Florida in the United States to Colombia and Venezuela. Within its range, three other crocodiles occur in freshwater

always visible. The scales on the underside of the body have sensory pits, seen as a small dot in the middle of each scale, even in tanned skins.

SALT SOLUTIONS

Like many vertebrates living in salt water, crocodiles must deal with extremes: too much salt and too little fresh water in their environment. In order to maintain the correct salt balance in their body fluids, they must excrete the excess salt they absorb directly through their skin and that they ingest while eating. However, reptiles cannot excrete sufficient amounts of salt unless their kidneys are flushed by large amounts of fresh water—a major hurdle for an animal living in an estuarine environment.

Crocodiles and other marine reptiles have overcome this problem by having glands capable of excreting high concentrations of salt without the loss of valuable water. The salt glands of marine turtles are modified tear glands, while the Galapagos Island's marine iguana expels a salty solution from glands in its nostrils. Crocodylids have salt glands on their tongue. These glands, which are actually modified salivary glands, secrete drops of salty solution. The crocodile's mouth is sealed off from its internal organs by a fleshy cartilaginous valve at the back of the throat, effectively making the tongue part of the external body surface.

This special method of excretion affords

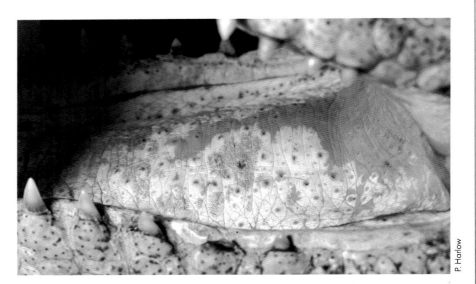

P. Harlow

crocodylids a high tolerance to salt water, and may explain their extensive distribution throughout the tropics. The lack of salt glands in the American alligator and in caimans (both members of the family Alligatoridae) may be the reason why these species have never successfully invaded salt water.

▲ A close-up of a crocodile's tongue shows the scattered, pore-like openings of the salt glands. These modified salivary glands excrete excess salt as a highly saline fluid to maintain an appropriate salt/water balance in the crocodile's body.

habitats: Morelet's crocodile *C. moreletii* in Mexico, Belize, and Guatemala; the Cuban crocodile *C. rhombifer* in Cuba; and the Orinoco crocodile *C. intermedius* in Venezuela and Colombia.

The wide-ranging Nile crocodile *C. niloticus* occurs throughout Africa (except for the waterless areas of the Sahara) and in Madagascar; this species occurs in estuaries, and in the interior of the continent it also inhabits rivers, swamps, and lakes. The only other member of the genus that occurs on the continent, the African slender-snouted crocodile *C. cataphractus*, is restricted to heavily forested areas of West and Central Africa. The ecological separation of this species and the Nile crocodile has not been studied.

The only crocodylid that does not belong to the genus *Crocodylus* is the dwarf crocodile *Osteolaemus tetraspis*. This tiny species is superficially similar to the dwarf caimans, and possibly its ecology in the heavily forested areas of West Africa is similar to that of the dwarf caimans in South America's Amazon basin. Virtually nothing is known of its life history or population dynamics.

THE GHARIAL FAMILY

The family Gavialidae contains two species of extremely thin-snouted crocodilians. Both are large. The false gharial *Tomistoma schlegelii* grows to lengths in excess of 4 meters (13 feet) and for the gharial *Gavialis gangeticus* lengths of of 6.5 meters (21 feet) have been recorded. The false gharial is classified in this family mainly from biochemical evidence; some authorities place it in the family Crocodylidae.

The extremely thin snouts of gharials are generally considered to be adaptations to a diet of fish, and it is unlikely that the snout of the true gharial could support the stresses involved in the capture of mammals as large as those taken by other crocodilians of similar size. Very thin snouts are also probably ineffective in habitats with thick obstructive vegetation.

The false gharial occurs on the Malay Peninsula of Thailand and Malaysia, and on the islands of Sumatra, Borneo, Java, and possibly Sulawesi (Celebes). It is distinctly marked with dark blotches, and adults are almost as colorful as juveniles. It is found in freshwater swamps, lakes, and rivers, but no detailed studies have been made of its ecology. The true gharial is one of the most distinctive crocodilians. The head with its thin snout and the weak legs appear disproportionately small in relation to the large body. Adult males develop a large fleshy knob on the tip of the snout that apparently serves to modify the sounds they make during social interactions. It spends more time in the water than most other crocodilians and frequents relatively fast-flowing reaches of the Indus, Bhima, Mahanadi, Ganges, Brahmaputra, Kaladan, and Irrawaddy rivers in Pakistan, India, Nepal, Bhutan,

▲ *Long thought to resemble the Indian gharial through convergent evolution only, recent biochemical studies have placed the Malayan or false gharial* Tomistoma schlegelii *in the same family (Gavialidae).*

▼ *The gharial* Gavialis gangeticus *of northern India is primarily a fish-eater, as suggested by its long slender snout. As males are usually characterized by a conspicuous knob at the tip of the snout, this is probably a female or subadult male. This species may grow to more than 6 meters (19 ¾ feet) in length.*

Jean-Paul Ferrero/AUSCAPE International

229

M.P. Price/Bruce Coleman Limited

▲ *Baby gharials hatching in Nepal. A solicitous parent, the female gharial buries her clutch of 30 to 50 eggs in a cavity dug in sand, then remains in the vicinity until they hatch 60 to 80 days later, digging out the nest and helping the young to break free of the eggshells.*

▼ *An Orinoco crocodile Crocodylus intermedius swallows a capybara, a pig-sized South American rodent. Rare and little-known, the Orinoco crocodile has a restricted distribution in Venezuela and Colombia.*

Bangladesh, and Burma. Its habitat has a high human population density, and net-fishing, dams, and egg collecting have seriously depleted populations. Fortunately, an intensive management and restocking program by the Indian government seems to be bringing the species back from the edge of extinction.

FOOD

Crocodilians eat a wide range of foods and generally adapt to what is most available in their habitat. Hatchlings eat mainly insects and other small invertebrates, then as they grow they eat more shrimp, snails, or crabs, depending on availability. Most species generally change over to fish as they approach maturity, but they will also take mammals, birds, and other vertebrates that live around the water's edge. Large adults may specialize on mammals, and Nile crocodiles have been seen to kill full-grown Cape buffaloes. Despite popular opinion, cannibalism is relatively rare among crocodilians and can usually be traced to social rather than nutritional causes.

Although they are master hunters in the aquatic arena, their large bodies are not very maneuverable on land, and potential prey are generally safe if more than about 4 meters (13 feet) from the water's edge. Humans are particularly clumsy mammals which spend a lot of time around the water, but fortunately for our species, most crocodilians do not usually regard us as suitable food items. Large crocodilians such as the American alligator, the black caiman, and the Indian mugger occasionally kill humans in defense of nests or territories, or for food, but such attacks are very rare considering the frequency of associations between these species

and high densities of people. In contrast, two species, the Indo-Pacific crocodile and the Nile crocodile, habitually hunt humans, and extreme care is necessary when near the water within the ranges of these species.

REPRODUCTION

All crocodilians lay eggs. The clutches of the smallest species average 10 to 15 eggs, while the Indo-Pacific crocodile may lay more than 50. All of the alligatorids, the false gharial and most crocodiles make nests of vegetation, soil, and debris, scraped up into a mound. The gharial and five species of crocodiles bury their eggs in holes in sandy beaches or other friable soil. The American crocodile sometimes makes mounds of sand in which it buries its eggs instead of burying them directly in the beach.

Most species do not simply abandon their eggs after laying. The mother, and sometimes the father, remain in the vicinity of the nest to defend it against predators. If the predator is a human or a bear, this nest defense can result in the death of the mother. Incubation takes 60 to 100 days, depending on the species and the incubation temperature, then the young crocodiles pip through the eggs and begin to call within the nest. In response to this, the adult will scrape a hole in the nest to release the babies. Often, roots or termite workings will have completely encased the eggs, so help from a parent is necessary for the babies to escape from the nest. Once the hatchlings are free of the nest, the adult will gently pick them up in its mouth and carry them to water. This incredibly delicate feat by a mother that weighs thousands of times more than her offspring was not generally believed possible until

David MacDonald/Oxford Scientific Films

filmed in the 1970s. A Nile crocodile has even been seen to pick up an unhatched egg from which the baby was calling and then roll it between the tongue and the roof of her mouth until the shell broke. The mother gently lowered the baby into the water before swallowing the empty egg shell.

Whether mounds of vegetation or holes in beaches, nests of crocodilians function as precise incubators. In the case of hole nests, the overburden of soil acts as a heat reservoir which dampens out the extreme temperature fluctuations at the surface. In mound nests, the temperature is maintained not only by the sun's heat but also from decaying vegetable material in the mound. In all nests, metabolic heat generated by the embryos raises the temperature by several degrees in the latter part of incubation. Whether they live in Himalayan mountain streams, or tropical jungles, or temperate swamps, all species of crocodilians incubate their eggs at about 30°C (86°F). Prolonged exposure to temperatures below about 27°C (81°F) or above about 34°C (93°F) kills embryos of most species.

The temperature at which the eggs are incubated is not only important for the survival of the embryos; it also determines the sex of the offspring. Unlike birds and mammals, the sex of whose embryos are determined at the moment of fertilization, the embryo in a newly laid crocodilian egg has no gender. The temperature at which the egg is incubated during the first few weeks determines whether the embryo will develop into a female or a male.

In the American alligator, Schneider's dwarf caiman, and probably the common caiman, temperatures less than about 31°C (88°F) produce females, and temperatures higher than about 32°C (90°F) produce males; intermediate temperatures produce both sexes. In some crocodylids, low temperatures produce all females, and intermediate temperatures, 31° to 33°C (88° to 91°F) produce males and females; however, higher temperatures again produce mainly females. The temperature of incubation has also been shown to affect color patterns and preferred temperatures of the offspring.

It has been suggested that the link between sex and temperature may be adaptive because of differences between the sexes in mating systems. Males of most species of crocodilians have strong social hierarchies, and it is probably only the largest that get to breed. It is therefore wasteful for a female to produce male offspring at less than optimal temperatures because this will result in low rates of growth after hatching. In contrast, all females will probably have a chance to breed, irrespective of their final size. It therefore makes sense to produce males at optimal temperatures for growth and females at other temperatures. While this theory is plausible, much more

Edward Robinson/Oxford Scientific Films

research needs to be done on the implications of temperature-dependent sex determination.

SOCIAL BEHAVIOR

Crocodilian social behavior is more complex than that reported for any other reptiles. Several researchers have seen what may be cooperative hunting by crocodilians. However, the interpretation of these observations is difficult because the motivation for the behavior may have been merely a tendency to group together at a food resource while maintaining a minimum personal distance (the studies were made in the wild). In contrast, social behavior during reproduction can easily be studied in captivity, and the motivation for the behavior is more obvious. Each species varies in its degree of sociality. Some, such as Schneider's dwarf caiman, live in territories from which they generally exclude adult members of their own sex. Others, such as some populations of the Nile crocodile, undertake seasonal migrations to nesting beaches where each male sets up a temporary territory for his harem of breeding females.

A crocodilian's jaws are terrible weapons, and adults are quite capable of killing each other. However, this rarely happens. The loser in a stand-off can indicate submission by vocalizations if it is a juvenile, or by lifting its head vertically to expose its throat. It is usually then allowed to slink off without further aggression. Bites, when they occur, tend to be on the base of the tail, and not on the relatively vulnerable chest region.

▲ *Throat bulging, an American alligator bellows at its neighbors in Everglades National Park, Florida. Perhaps the noisiest of the crocodilians, both sexes of the American alligator have an extraordinarily varied vocabulary. Territorial males utter loud bellowing notes, easily audible at 150 meters (nearly 170 yards), and repeated about every ten seconds; neighbors respond in choruses that may last half an hour or more. Adults also utter a range of grunting and hissing sounds, and courting couples exchange quiet cough-like notes.*

▲ A crèche of Indo-Pacific crocodiles. For the first few weeks after hatching, baby crocodiles remain in groups close to the water's edge, guarded by the mother and sometimes other adults, feeding on insects and other small aquatic life, and diving instantly at any sign of danger.

David Curl

Social signals include body postures such as the submission signal, head slaps on the water, grunts and roars that are audible to human listeners, and subaudible low-frequency sounds for communication under water. Most of these have been studied only in relation to reproduction, but they may also have more subtle everyday uses. Adults and subadults will respond to the distress calls of juveniles and defend them against predators whether they are related or not. In one instance, three researchers had their inflatable boat sunk by an adult Schneider's dwarf caiman when they imitated the distress calls of a hatchling common caiman. Hatchlings will often stay together in a group, accompanied by the mother for periods ranging from weeks to years, depending on the habitat and the species. During this period the hatchlings communicate among themselves and with the mother by means of audible grunts.

ECONOMIC EXPLOITATION

Crocodilian leather is fashionable, and products made from crocodilian skins fetch high prices. The most sought-after leather is from "classic skins" such as the Indo-Pacific crocodile, the American alligator, and the Nile crocodile. Belly skins of these species have few osteoderms (bony buttons below each scale), and the surface polishes to a glossy sheen. However, in recent years the increasing scarcity of classic skins has led to the use of skins of the common caiman. Wild specimens of this species have thick osteoderms, and the leather cannot be given a glossy sheen unless coated artificially.

Unrestricted hunting during the 1950s and 60s reduced most species with classic skins to close to the point of economic extinction. National and international legislation was enacted during the late 1960s and early 70s by many producer countries, and the focus moved to conservation. The success of those conservation actions is reflected in the fact that controlled economic exploitation is now allowed in most of the countries that initially enacted complete protection for their crocodilians. The countries that led the way to controlled economic exploitation were Papua New Guinea for the Indo-Pacific crocodile, the United States for the American alligator, Zimbabwe for the Nile crocodile and Venezuela for the common caiman. The exact form of management varies from country to country but most have a mix of the following three strategies.

Hunting

Adult and subadult crocodilians are hunted in the wild, but there is usually some legal limit on the the methods that can be used and the size that can be taken. In Papua New Guinea, only crocodiles above a certain size limit are protected. The smallest individuals are not hunted because the skins are too small for the trade. Thus the wild breeding stock of large, old crocodiles is protected and only an intermediate size group exploited. Presumably enough of these escape to grow and replenish the older age group. Populations of crocodilians that have been subjected to size-selective harvesting in Papua New Guinea, the United States, and Venezuela seem to be doing quite well. Although it might appear that hunting is a bad thing for a species, this is not necessarily so; crocodilians have shown themselves resistant to intense persecution if their habitats are maintained intact. In developing countries one of the most compelling reasons to maintain habitats is that the local people depend on these habitats economically. Hunting maintains the economic incentive and, paradoxically, in many areas is seen as an important conservation tool.

Ranching

Crocodilian ranching involves the collection of eggs or hatchlings (occasionally juveniles) in the wild for rearing in captivity. Growth rates in captivity, with the appropriate temperature and feeding regimes, are far higher than those attained in the wild, and the animals are usually slaughtered at between one and three years of age. Raising the animals in captivity usually results in better-quality, uniform-grade skins that achieve higher prices. It is also easier to control the conditions of slaughter for the production of the meat, which is increasingly becoming a restaurant delicacy. In many cases it is also easier for the official wildlife agency to monitor the exploitation because all killing takes place at the rearing stations. The collection of eggs or hatchlings in the wild also gives people an economic incentive to maintain natural habitats.

Farming

The breeding and raising of crocodilians on closed-cycle farms was the least-used method of exploiting crocodilians, but it has become more important as natural habitats are destroyed. It is potentially possible to select individuals for domesticated races of crocodilians that produce more eggs in captivity, grow faster, and have more fashionable skins. Research along these lines is being undertaken in the United States and Zimbabwe. Such activities are essentially irrelevant to the conservation of natural species and their habitats. However, farming has recently shown that it can affect conservation. The shifting of species of crocodilians around the world for farming operations poses a grave threat to many aquatic systems. Should the exotic species escape, the local crocodilians may be wiped out by exotic diseases or through competition with and predation by the introduced species. Many aquatic animals that have evolved to coexist with one species of crocodilian may not be able to survive if another is suddenly thrust upon them. It would be ironic if captive breeding, which was instrumental in the recovery of some species of crocodilians, should change the status of these species from threatened to threatening.

WILLIAM E. MAGNUSSON

LOCOMOTION

Crocodiles swim with lateral strokes of the tail which propel the body slowly but apparently effortlessly through the water. This type of locomotion must be very efficient, because some individuals have been seen at sea hundreds of kilometers from land. Crocodiles swim fast only to escape from danger or to pursue prey. The tail is capable of propelling the whole body vertically out of the water to snap prey from overhanging branches or to power the thrust as a large crocodile explodes out of the water to take a mammal that has ventured too close to the bank.

On land a crocodilian's tail is useless, and it must rely on its short legs for locomotion. Although superficially lizard-like, the body is too heavy to be pushed along the ground in a normal splayed-leg lizard-run for any distance. Crocodilians generally "belly slide" only if they are on mud or some other slippery surface, or if hard-pressed by a predator.

For long-distance or leisurely movements over dry land, crocodilians use a "high walk" similar to that of a walking mammal, with the legs positioned almost under the body and the belly held well off the ground. Some species, such as the common caiman *Caiman crocodilus*, regularly cover long distances over land. However, movement is slow, and although they may eat carrion it is unlikely that they can catch prey away from the water.

The most spectacular crocodilian gait is that of Johnston's crocodile *Crocodylus johnsoni*. To escape back to water over short distances, it gallops using the same stride sequence as a horse at high speed. The hind limbs propel the body forward and high off the ground. In fact, all four limbs may be off the ground at the same time. The forelimbs take the weight of the body as it lands, the body is flexed, and the hind limbs are swung forward to provide the next forward thrust. This method of escape can be used only by small crocodilians, and the belly slide is probably more effective unless the ground is firm or is covered by obstacles such as fallen logs.

John Carnemolla/Australian Picture Library

▲ As predators, crocodiles rely heavily on their terrifying ability to explode into sudden violent activity, completely unexpectedly—to humans at least— in such otherwise lethargic-seeming creatures. The sudden lunge of a crocodile can lift it almost entirely clear of the water.

Gunther Deichmann/AUSCAPE International

◀ Except for brief charges, crocodiles are generally slow and relatively awkward on land. A conspicuous exception is Johnston's crocodile *Crocodylus johnsoni* of Australia, which is capable of a full gallop in startled sprints back to the safety of the water.

FURTHER READING

- **INTRODUCING REPTILES & AMPHIBIANS (P. 14)**

Bellairs, A., 1969. *The Life of Reptiles.* Weidenfeld and Nicholson, London. 2 vols, 590 pp.

Duellman, W. E., & L. Trueb, 1994. *Biology of Amphibians.* Johns Hopkins University Press, Baltimore, Maryland. xvii + 670 pp.

Gans, C. (ed.), 1969 et seq. *Biology of the Reptilia.* Vol. 1 (1969) - Vol. 13 (1982), Academic Press, London; Vol. 14 (1985) - Vol. 15 (1985), John Wiley, New York; Vol. 16 (1988), A. Liss, New York; Vol. 17 (1992) & Vol. 18 (1992), University of Chicago Press, Chicago. (Continuing series.)

Halliday, T., & K. Adler (eds), 1986. *The Encyclopedia of Reptiles and Amphibians.* Facts on File, New York. xvi + 143 pp.

Pough, F.H., R.M. Andrews, J.E. Cadle, M.L. Crump, A.H. Savitzky & K.D. Wells, 1998. *Herpetology.* Prentice Hall, Upper Saddle River, NJ. xi + 577 pp.

- **CLASSIFYING REPTILES & AMPHIBIANS (P. 19)**

Duellman, W.E., 1993. *Amphibian species of the world: additions and corrections.* University of Kansas Museum of Natural History, special publication no. 21:1-372.

Frost, D.R, 1985. *Amphibian species of the world.* Allen Press and Assoc. of Systematic Collections, Lawrence, Kansas. v + 732 pp.

Sokolov, V.E., (ed.), 1988. *Dictionary of Animal Names in Five Languages. Amphibians and Reptiles.* Russky Yazyk Publishers, Moscow. 355 pp.

- **REPTILES & AMPHIBIANS THROUGH THE AGES (P. 24)**

Callaway, J.M., & E.L. Nicholls (eds), 1997. *Ancient Marine Reptiles.* Academic Press, San Diego. 501 pp.

Carroll, R.L., 1988. *Vertebrate Paleontology and Evolution.* W.H. Freeman and Co., New York. 698 pp.

Currie, P.J., & K. Padian, 1997. *Encyclopedia of Dinosaurs.* Academic Press, San Diego. 869pp.

Dingus, L., & T. Rowe, 1998. *The Mistaken Extinction. Dinosaur Evolution and the Origin of Birds.* W.H. Freeman and Co., New York. xiv + 332 pp.

Sumida, S.S., & K.L.M. Martin (eds), 1997. *Amniote Origins.* Academic Press, San Diego.

- **HABITATS & ADAPTATIONS (P. 30)**

Bradshaw, S.D., 1986. *Ecophysiology of Desert Reptiles.* Academic Press Australia, North Ryde. xxv+ 324 pp.

Duellman, W.E., & L. Trueb, 1994. *Biology of Amphibians.* Johns Hopkins University Press, Baltimore, Maryland. xvii + 670 pp.

Feder, M.E., & W.W. Burggren (eds), 1992. *Environmental Physiology of Amphibians.* University of Chicago Press, Chicago, Illinois.

Stebbins, R.C., & N.W. Cohen, 1995. *A Natural History of Amphibians.* Princeton University Press, Princeton. 316 pp.

Vitt, L.J., & E.R. Pianka, 1994. *Lizard Ecology.* Princeton University Press, Princeton. 403 pp.

- **REPTILE & AMPHIBIAN BEHAVIOR (P. 36)**

Carpenter, C.C., 1977. 'Communication in snakes', *American Zoologist,* 17:217-223.

Carpenter, C.C., 1978. 'Ritualistic social behavior in lizards', pp. 253-267 in N. Greenberg & P.D. MacLean (eds), *Behavior and Neurology of Lizards.* National Institutes of Mental Health.

Duellman, W.E., 1992. 'Reproductive strategies of frogs', *Scientific American,* 267:80-87.

Garrick, L.D., 1977. 'Social signals and behavior of adult alligators and crocodiles', *American Zoologist,* 17:225-239.

Greene, H.W., 1973. 'Defensive tail display by snakes and amphisbaenians', *Journal of Herpetology,* 7:143-161.

Heatwole, H., & B.K. Sullivan (eds), 1995. *Amphibian Biology. Vol. 2. Social Behaviour.* Surrey Beatty & Sons, Chipping Norton, Australia. pp. 419–710.

Jackson, C.G., & J.D. Davis, 1972. 'A quantitative study of the courtship display of the red-eared turtle *Chrysemys scripta elegans* (Weid)', *Herpetologica,* 28:58-64.

- **ENDANGERED SPECIES (P. 42)**

IUCN, 1996. *1996 IUCN Red List of Threatened Animals.* IUCN, Gland, Switzerland. 368 pp. (Also on http://www.iucn.org/themes/ssc/redlist/redlist.htm)

- **CAECILIANS (P. 52)**

Duellman, W.E., & L. Trueb, 1994. *Biology of Amphibians.* Johns Hopkins University Press, Baltimore, Maryland. xvii + 670 pp.

Nussbaum, R.A., 1977. 'Rhinatrematidae: A new family of caecilians (Amphibia: Gymnophiona)', *Occasional Papers of the Museum of Zoology, University of Michigan,* 683:1-30.

Nussbaum, R.A., 1979. 'The taxonomic status of the caecilian genus *Uraeotyphlus* Peters', *Occasional Papers of the Museum of Zoology, University of Michigan,* 687:1-20.

Nussbaum, R.A., 1983. 'The evolution of a unique jaw-closing mechanism in caecilians (Amphibia: Gymnophiona) and its bearing on caecilian ancestry', *Journal of Zoology, London,* 199:545-554.

Nussbaum, R.A., & M. Wilkinson, 1989. 'On the classification and phylogeny of caecilians (Amphibia: Gymnophiona), a critical review', *Herpetological Monographs,* 3:1-42.

Nussbaum, R.A., & M. Wilkinson, 1995. 'A new genus of lungless tetrapod: a radically divergent caecilian (Amphibia: Gymnophiona)', *Proceedings of the Royal Society, London,* 261: 331-335.

- **SALAMANDERS & NEWTS (P. 60)**

Conant, R., & J.T. Collins, 1991. *A field guide of the reptiles and amphibians of eastern and central North America.* Houghton Mifflin Company, Boston.

Duellman, W.E., & L. Trueb, 1994. *Biology of Amphibians.* Johns Hopkins University Press, Baltimore, Maryland.

Lanza, B., V. Caputo, G. Nascetti & L. Bullini, 1995. *Morphologic and genetic studies on the European plethodontid salamanders: taxonomic inferences (genus* Hydromantes). Museo Regionale di Scienze Naturali, Torino.

Rimpp, K., 1985. *Salamander und Molche.* Eugen Vemer GmbH & Co, Stuttgart.

Stebbins, R.C., 1985. *A field guide to western reptiles and amphibians. Field marks of all species in western North America.* Houghton Mifflin Company, Boston.

• FROGS & TOADS (P. 76)

Duellman, W.E., & L. Trueb, 1994. *Biology of Amphibians.* Johns Hopkins University Press, Baltimore, Maryland. xvii + 670pp.

Heatwole, H., & E.M. Dawley (eds), 1997. *Amphibian Biology. Vol. 3. Sensory Perception.* Surrey Beatty & Sons, Chipping Norton, Australia. pp. 711–972.

Mattison, C., 1987. *Frogs and Toads of the World.* Facts on File, New York. 191 pp.

Stebbins, R.C., & N.W. Cohen, 1995. *A natural history of amphibians.* Princeton University Press, Princeton. xvi + 316 pp.

Vial, J.L., (ed.), 1973. *The Evolutionary Biology of Anurans.* University of Missouri Press, Columbia. 470 pp.

• TURTLES & TORTOISES (P. 108)

Bjorndal, K.A., (ed.), 1995. *Biology and Conservation of Sea Turtles.* Rev. Ed. Smithsonian Institution Press, Washington, D.C. 615 pp.

Ernst, C.H., & R.W. Barbour, 1989. *Turtles of the World.* Smithsonian Institution Press, Washington, D.C. xii + 313 pp.

Lutz, P.L., & J.A. Musick, 1997. *The Biology of Sea Turtles.* CRC Press, Boca Raton, Florida. 432 pp.

Obst, F.J., 1985. *Die Welt der Schildkroten.* Edition Leipzig, Leipzig. 235 pp.

Pritchard, P.C.H., 1979. *Encyclopedia of turtles.* T.F.H. Publications, Neptune, NJ. 895 pp.

• LIZARDS (P. 126)

Greer, A.E., 1989. *The Biology and Evolution of Australian Lizards.* Surrey Beatty & Sons, Chipping Norton, Australia. xvi + 264 pp.

Huey, R.B., E.R. Pianka & T.W. Schoener (eds), 1983. *Lizard Ecology: Studies of a Model Organism.* Harvard University Press, Cambridge. 501 pp.

Mattison, C., 1989. *Lizards of the World.* Facts on File, Inc., New York. 192 pp.

Pianka, E.R., 1986. *Ecology and Natural History of Desert Lizards.* Princeton University Press, Princeton. x + 208 pp.

Smith, H.M., 1946 (1995 edn). *Handbook of Lizards - Lizards of the United States and Canada.* Cornell University Press, Ithaca. xxxi + 557 pp.

Vitt, L.J., & E.R. Pianka (eds), 1994. *Lizard Ecology - Historical and Experimental Perspectives.* Princeton University Press, Princeton. xii + 403 pp.

• SNAKES (P. 174)

Campbell, J.A., & W.W. Lamar, 1989. *The Venomous Reptiles of Latin America.* Cornell University Press, Ithaca, New York. xii + 425 pp.

Engelman, W.-E., & F.J. Obst, 1981. *Snakes. Biology, behavior and relationship to man.* Edition Leipzig, Leipzig. 222 pp.

Greene, H.W., 1997. *Snakes: The Evolution of Mystery in Nature.* California University Press, Berkeley. 351 pp.

Mattison, C., 1987. *Snakes of the World.* Facts on File, New York. 191 pp.

Mehrtens, J.M., 1987. *Living Snakes of the World.* Sterling Publishing Co., Inc., New York. 480 pp.

Seigel, R.A., J.T. Collins & S.S. Novak (eds), 1987. *Snakes: Ecology and Evolutionary Biology.* Macmillan, New York. 529 pp.

Seigel, R.A., & J.T. Collins, 1993. *Snakes. Ecology and Behavior.* McGraw-Hill, New York.

Shine, R., 1994. *Australian Snakes: A Natural History.* Cornell University Press, Ithaca, NY, and Reed Books, Sydney, NSW.

• AMPHISBAENIANS (P. 212)

Broadley, D.G., 1997. 'A review of the *Monopeltis capensis* complex in southern Africa (Reptilia: Amphisbaenidae)', *African Journal of Herpetology,* 46(1):1-12.

Gans, C., 1990. 'Patterns in amphisbaenian biogeography. A preliminary analysis', pp. 133-143 in G. Peters & R. Hutterer (eds), *Vertebrates in the Tropics.* Museum Alexander Koenig, Bonn.

• TUATARA (P. 218)

Cree, A., & D. Butler, 1993. *Tuatara recovery plan* (Sphenodon *spp.*). Threatened Species Recovery Plan Series No. 9. New Zealand Department of Conservation, Wellington.

Cree, A., C.H. Daugherty & J.M. Hay, 1995. 'Reproduction of a rare New Zealand reptile, the tuatara *Sphenodon punctatus,* on rat-free and rat-inhabited islands', *Conservation Biology,* 9: 373-383.

Cree, A., M.B. Thompson & C.H. Daugherty, 1995. 'Tuatara sex determination', *Nature,* 375: 543.

Daugherty, C.H., A. Cree, J.M. Hay & M.B. Thompson, 1990. 'Neglected taxonomy and continuing extinctions of tuatara *(Sphenodon)', Nature,* 247: 177-179.

Newman, D.G., P.R. Watson & I. McFadden, 1994. 'Egg production by tuatara on Lady Alice and Stephens Island, New Zealand', *New Zealand Journal of Zoology* 21: 387-398.

• CROCODILES & ALLIGATORS (P. 224)

'Biology of the Crocodilia', *American Zoologist,* 29(3):823-1054. (Special symposium with 17 papers on various aspects of crocodilian biology).

Guggisberg, C.A.W., 1972. *Crocodiles: Their Natural History, Folklore and Conservation.* David and Charles, Newton Abbot. 204 pp.

Ross, C.A., & S. Garnett (eds), 1989. *Crocodiles and Alligators.* Facts on File, New York. 240 pp.

Webb, G.J.W., S.C. Manolis & P.J. Whitehead (eds), 1987. *Wildlife Management: Crocodiles and Alligators.* Surrey Beatty & Sons, Chipping Norton. 552 pp.

INDEX

ACKNOWLEDGMENTS

The editors and publishers would like to thank the following people for their assistance: Elizabeth Cameron and Sophia Zantiotis, Australian Museum; Rod Scott, Big Nose Bros. Creative Communications; Beverley Barnes; Tristan Phillips; Alison Pressley; Lu Sierra; Tracy Tucker; and Natalie Vellis.